Lecture Notes in Computer Science   12620

More information about this subseries at http://www.springer.com/series/8183

Marina L. Gavrilova · C. J. Kenneth Tan (Eds.)

Transactions on Computational Science XXXVIII

 Springer

Editors-in-Chief
Marina L. Gavrilova
University of Calgary
Calgary, AB, Canada

C. J. Kenneth Tan
Sardina Systems OÜ
Tallinn, Estonia

ISSN 0302-9743   ISSN 1611-3349   (electronic)
Lecture Notes in Computer Science
ISSN 1866-4733   ISSN 1866-4741   (electronic)
Transactions on Computational Science
ISBN 978-3-662-63169-0   ISBN 978-3-662-63170-6   (eBook)
https://doi.org/10.1007/978-3-662-63170-6

This Springer imprint is published by the registered company Springer-Verlag GmbH, DE
part of Springer Nature
The registered company address is: Heidelberger Platz 3, 14197 Berlin, Germany

LNCS Transactions on Computational Science

Computational science, an emerging and increasingly vital field, is now widely recognized as an integral part of scientific and technical investigations, affecting researchers and practitioners in areas ranging from aerospace and automotive research to biochemistry, electronics, geosciences, mathematics, and physics. Computer systems research and the exploitation of applied research naturally complement each other. The increased complexity of many challenges in computational science demands the use of supercomputing, parallel processing, sophisticated algorithms, and advanced system software and architecture. It is therefore invaluable to have input by systems research experts in applied computational science research.

Transactions on Computational Science focuses on original high-quality research in the realm of computational science in parallel and distributed environments, also encompassing the underlying theoretical foundations and the applications of large-scale computation.

The journal offers practitioners and researchers the opportunity to share computational techniques and solutions in this area, to identify new issues, and to shape future directions for research, and it enables industrial users to apply leading-edge, large-scale, high-performance computational methods.

In addition to addressing various research and application issues, the journal aims to present material that is validated – crucial to the application and advancement of the research conducted in academic and industrial settings. In this spirit, the journal focuses on publications that present results and computational techniques that are verifiable.

Scope

The scope of the journal includes, but is not limited to, the following computational methods and applications:

- Aeronautics and Aerospace
- Astrophysics
- Big Data Analytics
- Bioinformatics
- Biometric Technologies
- Climate and Weather Modeling
- Communication and Data Networks
- Compilers and Operating Systems
- Computer Graphics
- Computational Biology
- Computational Chemistry
- Computational Finance and Econometrics
- Computational Fluid Dynamics

- Computational Geometry
- Computational Number Theory
- Data Representation and Storage
- Data Mining and Data Warehousing
- Information and Online Security
- Grid Computing
- Hardware/Software Co-design
- High-Performance Computing
- Image and Video Processing
- Information Systems
- Information Retrieval
- Modeling and Simulations
- Mobile Computing
- Numerical and Scientific Computing
- Parallel and Distributed Computing
- Robotics and Navigation
- Supercomputing
- System-on-Chip Design and Engineering
- Virtual Reality and Cyberworlds
- Visualization

Editorial

The *Transactions on Computational Science* journal is published as part of the Springer series *Lecture Notes in Computer Science*, and is devoted to a range of computational science issues, from theoretical aspects to application-dependent studies and the validation of emerging technologies.

The journal focuses on original high-quality research in the realm of computational science in parallel and distributed environments, encompassing the theoretical foundations and the applications of large-scale computations and massive data processing. Practitioners and researchers share computational techniques and solutions in the area, identify new issues, and shape future directions for research, as well as enable industrial users to apply the presented techniques.

The current issue is devoted to research on modelling, optimization and graphs, with applications in 3D and sketch modelling, engineering design, evolutionary computing and networks.

The first article of the issue "Efficient Creation of 3D Organic Models from Sketches and ODE-based Deformations" introduces an efficient way to create organic 3D models from sketches. An ordinary differential equation and sketch-guided deformation algorithm is proposed to ensure an exact fit for the two corresponding sketch segments. The research is validated on a collection of sketches and establishes that the method can create 3D models easily and efficiently.

The second article "Tree Species Modelling for Digital Twin Cities" delivers a new approach to 3D tree modeling for virtual city planning. It presents a way for users to create vegetation content for a digital twin city through dynamic 3D plant models on a large scale. Based on the given species profile and solving for the unknowns within the growth space constraints, a species model is developed through formulated growth rules. The authors demonstrate that this approach produces structurally representative species models with respect to their actual physical and species characteristics.

The third article "Self-restructuring of Mesh-connected Processor Arrays with Spares Assigned on Rotated Orthogonal Side" addresses the problem of processing a vast amount of information in real time and near-real time. It presents a method for enhancing the run-time reliability of a processing array in mission critical systems. The main advantage of the method is that it operates without an external host computer or manual maintenance operations.

The fourth article "Algorithms for Generating Strongly Chordal Graphs" presents an elegant solution to the strongly chordal graph generation problem. Graph generation can be used in cataloging and testing conjectures. Strongly chordal graphs are a subclass of chordal graphs for which polynomial-time algorithms can be designed for problems which are NP-complete for the parent class. The authors propose three different algorithms for generating strongly chordal graphs, each based on a different characterization.

The fifth article "A New Bio-heuristic Hybrid Optimization for Constrained Continuous Problems" presents a new novel bio-inspired evolutionary algorithm. The method is based on the hybrid amalgamation of two nature-inspired methods and uses the notion of elitism as an important characteristic of evolutionary algorithms. Extensive experiments demonstrate that the solutions of the constrained optimization problems found by using the proposed method are highly accurate.

The sixth article "Analysis and Optimization of Low Power Wide Area IoT Network" continues the topic of optimizaton, but with application to the Internet of Things. Particularly, the article proposes a method that can enhance IoT applications because of its extended battery life, low data rate and large coverage area. The paper also investigates the effects of different transmission parameters of long-range networks on the battery life of sensors.

The seventh article "Novel Hybrid GWO-WOA and BAT-PSO Algorithms for Solving Design Optimization Problems" presents a mechanism of hybridization to enhance the performance of two hybrid nature-inspired algorithms. The algorithms studied are the two nature-inspired algorithms based on Grey Wolf Optimizer and Whale Optimization Algorithm and Binary Bat Optimization Algorithm and Particle Swarm Optimization Algorithm. Experimentation confirms that the proposed algorithms are capable of solving constrained complex problems with diverse search spaces.

We thank all of the reviewers for their diligence in making recommendations and evaluating revised versions of the papers presented in this journal issue. We would also like to thank all of the authors for submitting their papers to the journal and the associate editors for their valuable work.

It is our hope that this collection of seven articles presented in this special issue will be a valuable resource for Transactions on Computational Science readers and will stimulate further research in the vibrant area of computational science theory and applications.

January 2021

Marina L. Gavrilova
C. J. Kenneth Tan

LNCS Transactions on Computational Science – Editorial Board

Contents

Efficient Creation of 3D Organic Models from Sketches and ODE-Based Deformations

Ouwen Li[1]($\boxtimes$), Shaojun Bian[1], Algirdas Noreika[2], Ismail Khalid Kazmi[3], Lihua You[1], and Jianjun Zhang[1]

[1] National Centre for Computer Animation, Bournemouth University, Dorset, UK
oli@bournemouth.ac.uk
[2] Indeform Ltd., K. Petrausko g. 26, 44156 Kaunas, Lithuania
[3] Teesside University, Middlesbrough, Tees Valley, UK

Abstract. Efficient creation of 3D organic models is an important topic. In this paper, we propose a new approach to create rough 3D organic base models easily and quickly. The proposed approach first generates 2D sketches by manually drawing outlines of 3D organic objects or extracting outlines from 2D images. Then the generated outlines are decomposed into parts. For each of the decomposed parts, a central line is calculated from the two corresponding sketch segments. A straight cylinder is bent so that its central line coincides with the central line of the two corresponding sketch segments. The radius of the cylinder is determined by minimizing the sum of the squared errors between the projected silhouette curves of the placed cylinder and the two corresponding sketch segments. Finally, an ordinary differential equation (ODE)-based and sketch-guided deformation algorithm is proposed to deform the cylinder so that the projected silhouette curves of the deformed cylinder exactly fit the two corresponding sketch segments. The experiment carried out in this paper demonstrates that the approach proposed in this paper can create 3D organic base models from 2D sketches easily and quickly.

Keywords: 3D organic base model creation · Sketch generation · Automatic placements of 3D cylinders · ODE-based deformations

1 Introduction

Polygon, NURBS and subdivision are three mainstream modeling approaches with applications in geometric modeling, computer graphics, computer animation and computer-aided design etc. They can create detailed and realistic 3D organic models and have been widely applied in various software packages such as Autodesk Maya and 3DS Max. With these approaches, modellers usually import reference images from the concept artists, and then begin the modeling and sculpting operations. These reference images, sometimes called model sheets, are the profiles from T-pose or A-pose in front view, side view and back view

© Springer-Verlag GmbH Germany, part of Springer Nature 2021
M. L. Gavrilova and C. J. K. Tan (Eds.): Trans. on Comput. Sci. XXXVIII, LNCS 12620, pp. 1–16, 2021.
https://doi.org/10.1007/978-3-662-63170-6_1

respectively, or various arbitrary poses in arbitrary perspective views. Although these modeling approaches are very powerful and popular, they require great effort to learn how to use them, involve heavy manual operations, and take a lot of time to complete modeling tasks.

In order to address the above problems, various sketch-based modeling (SBM) approaches have been developed in the past several decades [24]. These approaches use the silhouettes of 3D objects to describe 3D models. According to Hoffman and Singh [11], for human vision systems, silhouettes alone are efficient for recognising everyday objects, by indexing their shapes in human memory. Also, we noticed that the most natural way for both novice modellers and veteran modellers to generate a new shape is to sketch freehand strokes with paper-pen-like graphic input tablets. So does the shape modification. In addition, the recent surge of painting or modeling applications in virtual reality such as Gravity Sketch [10], Tilt Brush [9], Quill [7], Medium [23] and Mozilla A-Painter [20] also show artists have no problems in sketching in 3D canvas.

Once silhouettes of 3D objects are obtained, 3D models are created from these silhouettes. Various sketch-based modeling approaches have been developed. All of them can be roughly divided into: direct shape generation, template-based mesh creation, and sketch-based deformation. Direct shape generation approaches are to create 3D models directly from 2D sketches. They can be further classified into: primitive-based 3D model creation and surface inflation. The approaches of template-based mesh creation deform a template 3D model to fit 2D sketches. Sketch-based Deformation methods use sketches to guide deformations of 3D models. Among these three different types of approaches, most research studies focus on creating 3D models directly from 2D sketches. The research study described in this paper belongs to this type. It quickly creates rough 3D models directly from 2D sketches.

Physics-based geometric modeling can greatly improve realism. Unfortunately, they usually involve heavy numerical calculations. Such a weakness makes physics-based geometric modeling less ideal for real-time 3D model creation. Ordinary differential equation (ODE)-based geometric modeling and shape deformations use the solution to a vector-valued ordinary differential equation to create 3D models and achieve shape deformations. Ordinary Differential Equations (ODE) have been widely used to describe various physical laws in scientific computing and engineering applications. For example, fourth-order ODEs have been used to describe lateral bending of elastic beams in structural engineering. In this paper, we will use such a fourth-order ordinary differential equation to develop a new primitive deformation method.

A preliminary version of this work has been reported in [19]. It will first manually draw silhouette sketches of 3D organic objects or extract silhouette sketches from 2D images. Then the obtained silhouette sketches are decomposed into parts. For each of the decomposed parts, a central line is calculated from the two corresponding sketch segments. A straight cylinder is bent to make its central line coincide with the central line of the two corresponding sketch segments.

The radius of the cylinder is determined by minimizing the sum of the squared errors between the projected silhouette curves of the placed cylinder and the two corresponding sketch segments. Finally, an ordinary differential equation (ODE)-based and sketch-guided deformation algorithm is proposed to deform the placed cylinder. It makes the projected silhouette curves of the deformed cylinder exactly fit the two corresponding sketch segments. With the proposed approach, 3D organic base models are easily and quickly created.

The remaining parts of the paper will be structured below. First, the existing work will be reviewed in Sect. 2. Next, manually drawing 2D sketches and extracting the silhouette contours of 2D images will be discussed in Sect. 3. After that, automatic placements of 3D cylinders guided by 2D sketches will be examined in Sect. 4. How to develop ODE-based and sketch-guided primitive deformations and fitting 3D cylinders to 2D sketches will be investigated in Sect. 5. Finally, conclusions will be drawn and future work will be discussed in Sect. 6.

2   Related Work

Sketch-based modeling and ODE-based geometric creation and shape deformations are two cornerstones. In this section, we review the existing work in these two fields.

2.1   Sketch-Based Modeling

Over the past several decades, sketch-based-modeling (SBM) has been widely studied in the computer graphic community [24]. It can be divided into: direct shape generation, template-based mesh creation, and sketch-based deformation. Direct shape generation can be further classified into: primitive-based 3D model creation and surface inflation.

Direct Shape Generation. In the category of direct shape generation, several systems have been proposed to generate organic models. Here, we review two types of mesh: primitive-based 3D model creation and surface inflation.

Primitive-Based 3D Model Creation. An interesting phenomena is that many modeling system's building bricks are geometrical primitives. In order to understand why geometrical primitives are chosen as the starting point of modeling a figure, we must first know how human vision system understands figures and interprets their shapes. [11] presented a theory of part salience. The theory builds on the minima rule for defining part boundaries. According to this rule, human vision defines part boundaries at negative minima of curvature on silhouettes and along the negative minima of the principal curvatures on surfaces. They propose that the salience of a part depends on (at least) three factors: its size relative to the whole object, the degree to which it protrudes, and the strength of its boundaries. Based on this rule, a human character has many component parts which are often differ in their visual salience. The sizes of some parts like torso relative to the whole object are substantial. The parts like head, limbs and foot protrude much from the main shape. The parts like neck have boundaries of sharp curvature changes.

The evidences show that these factors influence visual processes which determine the choice of figure and ground. Hence, we divide human figure into different parts based on these factors. Primitives-based systems decompose the modeling task as a process of creating a certain set of geometry primitives and further editing the primitives [5,25,29]. The idea of assembling simple geometric primitives to form 3D models is very common in CSG (Constructive Solid Geometry) modeling. Shtof et al. [25] introduced a snapping method to detect the feature curves and silhouette curves of both 2D sketches and 3D simple geometric primitives, i.e. box, sphere, cylinder and cone. The method automatically determined the core parameters of these simple geometric primitives to fit the 2D sketches. Then, it improved the model globally by inferring geosemantic constraints defining the relationships between different parts. These relationships include: parallelism, orthogonality, collinear centers, concentric, and coplanar. In the work of [5], the researchers proposed a tool to generate a primitive from only 3 strokes. The first two strokes define the 2D profile and the last stroke defines the axis along which the profile curve sweeps. However, the work is only designed for man-made objects. Structured annotations for 2D-to-3D modeling [8], on the other hand, focus on organic modeling. It used two sets of primitives. One set of primitives are generalized cylinders. They are created by the input of a single open sketch stroke represented as a spline. Then, they are modified by using simple gestures such as tilting cross sections, scaling local radius, rotating symmetrical plane, and changing cap size. The other set of primitives are ellipsoids. They are generated according to the drawn closed ellipse sketch strokes. In the work, a set of annotation tools were developed from the semantic information such as symmetric, equal length and angle, and alignment to further edit the surface shape. Unfortunately, these annotation tools are unable to create complicated shapes.

Surface Inflation. Another way of generating a shape inflates a surface represented by a closed 2D region/sketch to give it a volume [15]. The surface inflation technique extrudes a polygonal mesh from a given skeleton outwards. It does a good job in modeling stuffed toys. One trend in surface inflation is to inflate free-form surfaces to create simple stuffed animals and other rotund objects in a SBM fashion [12,13,21]. The pioneering Teddy system [12] takes closed curves as inputs, finds their chordal axes as the spines, then wraps the spines with a polygonal mesh. Later on, [21] enriches editing operations for the inflated base mesh. This approach also presents two types of the control curves: smooth and sharp. A smooth curve constrains the surface to be smooth across it, while a sharp curve only places positional constraints with C^0 continuity. Sharp control curves appear when operations like cutting, extrusion, and tunnel take place. They also serve the creation of creases on the surface. The Smooth Sketch system [13] supports the creation of cusps and T-junctions, which Teddy and its successors fail to address. In addition, SmoothSketch [13] extended Teddy's work [12] to an extent that the strokes do not need to be closed. Although the surface inflation approach is good at creating stuffed animals and simple toys, it is not versatile enough to express geometrical details on the surface. BendSketch [18] offers a technique which enables complex curvature patterns existing on surfaces. In order to give the bending information, users need to draw a set of lines that comply with what the BendSketch system [18] has specified. This treatment mimics the hatching technique which artists often utilise to express the sense of volume and curvature information on the surfaces.

Template-Based Mesh Creation. Using geometric primitives to align 2D sketches provides an easy and efficient approach to generate rough base models. However, manipulating the generated rough base models in one image plane only is difficult to create detailed 3D models. For template-based mesh creation, an elegant technique for sketch-based modeling has been proposed in [17] to find precise correspondences and determine mesh deformations. By combining skeleton-based deformation and mesh editing, an efficient approach was proposed in [14] to quickly deform a 3D template model to fit the user's drawn sketches. Among various approaches of direct shape generation from user-drawn sketches, a variety of sketch-based modeling tools are based on a "sketch-rotate-sketch" workflow. Such a workflow requires users to draw sketches from many views, causing the difficulties in matching input strokes with a guidance image and finding a good view [8].

Sketch-Based Deformation. Deformation tools provide an interface for users to interact and modify the mesh surface. A good deformation tool should meet the following principles [2]:

- Flexibility: permit users to change the mesh surface as they wish, meanwhile reserve correct modeling constraints.
- Shape quality: aesthetically pleasing
- Intuitive result: conform to natural deformations happening in real world with real physical material, which makes no recognition confusion for users when they modify the shape.

Sketch-based shape editing like Teddy and others [12,13,21] features inflation approaches using smooth silhouettes. Therefore, they create only smooth shapes. In order to create aesthetically pleasing details on meshes, SBM should be equipped with shape editing methods supporting inserting feature curves while preserving the global and local geometry. Geometry-based editing system is based on the energy in a form of geometrical traits. A popular geometry-based frame is Laplacian/Poisson model, which represents the differential traits of the surface in various ways depending on how they are employed. In the discrete Laplacian/Poisson model, it is easy to displace a set of edges (e.g., sketch a new position of an identified contour) while preserve the geometric details of the surface as much as possible [22]. In order to address the problem that the differential properties of feature lines are related to the viewing direction, the feature lines should not coincide with edges on the mesh. [21] extends the framework of Laplace/Poisson mesh modeling in 3 ways. First, it accommodates constraints on the normals and the curvature. Second, it allows constraints to be placed on virtual vertices, i.e. vertices placed on edges that only serve to implement the constraints but are never added to the mesh. The virtual vertices are linearly interpolated on the edge between 2 vertices. Third, it incorporates a tangential mesh regularization, which moves edges onto sharp features while ensures well-shaped triangles. The method supports the functions of changing the moderate noisy silhouette contours, modifying the sharp feature lines like ridge, ravine and crease, and editing smooth features and suggestive contours. Apart from the deformation based on differential surface representations, another family makes use of multi-resolution or subdivision editing as a way to perform global surface deformations while preserve local surface details. Subdivision mesh is firstly decomposed into a low-frequency base surface and a high-frequency detail information. After deforming the base surface, the detail information is added to the deformed base surface. In the case of interactive shape editing by means of manipulating control handles with 3 degrees of freedom for translation and 3 degrees of freedom for rotation, the displacement vector is defined as the change of position and orientation of the control handles. The final shape can be seen as the outcome of adding a basis function of displacement vector to the origin shape. The requested displacement has to be translated into coefficients of this basis functions. These coefficients include but not limited to smoothness, stiffness (order of energy function), fullness, etc. Besides, the basis functions require the precomputation of an inverse matrix to speed up the surface updating [1], which indicates these approaches are very computational-expensive. [22] proposed a sketch-based mesh editing system that can deform a user selected ROI (region of interest) to fit the input sketch. [33] is able to compute from the oversketched feature line alone.

2.2   ODE-Based Geometric Creation and Shape Deformations

Compared to the geometry-based technology, physic-based editing technology simulates the physical principle through physical energy function. It complies with the real-world physical fidelity. However, it is less flexible if 3D artists want to achieve very drastic effects because the penalty of stretching or bending forces on the energy function will be very big. Another challenge of physically based skin deformation methods is that they are computational-expensive, far more inefficient than geometry based approaches. Hence, our object is to improve the performance and efficiency of physical-based editing scheme by decreasing the 2-dimensional partial differential equation to 1-dimensional ordinary differential equation. The finite element method and the finite difference method are widely used to solve various partial differential equations in this scenario. In this paper, we choose the finite difference method to solve the converted ordinary differential equation (ODE).

ODEs have been widely applied in scientific computing and engineering analyses to describe the underlying physics. For example, fourth-order ODEs have been used to describe lateral bending deformations of elastic beams. Introducing ODEs into geometric processing can create physically realistic appearances and deformations of 3D models. ODE-based sweeping surfaces [31], ODE-based surface deformations [4, 32], and ODE-based surface blending [30] have also been developed previously.

Although researchers studied ODE-based geometric surface creation and deformations, how to use ODE-based modeling to deform geometric primitives and create new shapes from the user's drawn sketches has been under-explored to date.

3 Generation of 2D Sketches

The first step of our proposed approach is to generate 2D sketches. Two methods are included in our developed system to generate 2D images. They are manually drawing without a reference image and extracting a 2D sketch from a reference image.

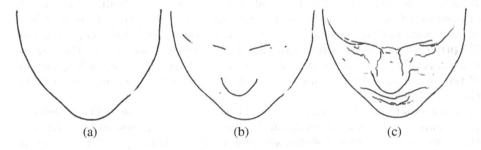

(a)   (b)   (c)

Fig. 1. Three types of sketches: (a) silhouette contour, (b) occluding contour, and (c) suggestive contour of a face model. (Image source: Wuhrer and Shu [28]).

As investigated by Wuhrer and Shu [28], three different types of sketches can be drawn by users. These three types of sketches are: silhouette contours, occluding contours, and suggestive contours (Fig. 1). Since this paper aims to develop a simple, easy, and efficient approach to create base 3D organic models from 2D sketches, we focus on silhouette contours only.

Our developed system allows users to draw the silhouette contours as shown in Fig. 2a. If a reference image such as the one give in Fig. 2b is available, our developed system allows users to trace the silhouette contours of the input image and obtain the silhouette contours of the input image as shown in Fig. 2c.

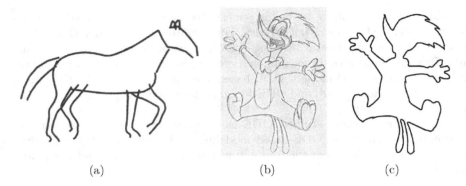

Fig. 2. Generation of silhouette contours: (a) user-drawn silhouette contour, (b) input 2D image, (c) silhouette contour generated from the input 2D image

4   Automatic Placements of 3D Primitives

The second step of our proposed approach is to select suitable 3D primitives and automatically place the 3D primitives to best fit the user-generated 2D sketches. It involves: decomposing a 2D sketch into parts with each part defined by a closed or open sketch segment or two open sketch segments, calculating the central line of each part, initially placing a primitive along the central line, projecting the 3D primitive to obtain one or two silhouette contours, and scaling the size of the placed 3D primitive to minimize the error between the silhouette contours of the placed 3D primitive and the sketch segments of the decomposed part as elaborated below.

For a female sketch shown in Fig. 3a, 11 parts are decomposed. The 11 decomposed parts are 1 head, 1 neck, 2 hands, 2 arms, 1 torso, 2 legs, and 2 ft. Among them, the head is a closed sketch segment, each of 2 hands and 2 ft is an open sketch segment, and all other parts are defined by two open sketch segments.

The study carried out in [16] combines three 3D primitives to create rough 3D models. The three 3D primitives are: generalized cylinders, ellipsoids, and cubes. Organic objects have no sharp edges. Cubes may not be suitable in representing organic models. Although generalized cylinders are powerful in mathematically representing various cylinder-like shapes, they are difficult to operate in practice since the mathematical equation defining a generalized cylinder involves a variable and all the parameters defining the shape and size of a generalized cylinder are dependent on the variable.

Based on the above discussions, this paper uses standard cylinders and ellipsoids as 3D primitives to represent 3D models. For ellipsoids, we scale them to best fit the 2D sketches like the female head which looks like an ellipse. For standard cylinders, we propose an ordinary differential equation (ODE)-based and sketch-guided deformation algorithm to change standard cylinders into various shapes to better approximate complicated 3D models. In what follows, we take

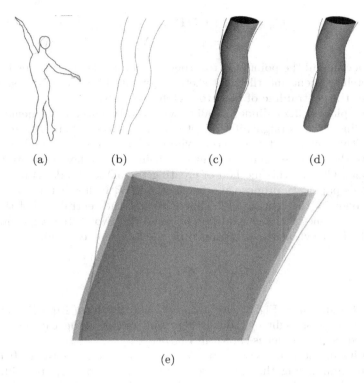

(a)   (b)   (c)   (d)

(e)

Fig. 3. Process of automatically placing curved cylinder

the left leg of the female sketch to demonstrate how to automatically place a cylinder to fit the sketch segments of the left leg.

As shown in Fig. 3b, the left leg is defined by a left sketch segment and a right sketch segment. In order to place a 3D cylinder to best align with the sketch segments, we first determine the central line of the left leg. There are different methods which can be used to determine a central line of two sketch segments. For example, we can generate a straight line from one starting point on one sketch segment and perpendicular to the sketch segment, find the intersecting point between the straight line and the other sketch segment, and use the middle point between the starting point and the intersecting point to create the central line. Here, we use the method below to generate central lines.

The total points on the left sketch segment may not be the same as the total points of the right sketch segment. From the points on each of the two sketch segments, we can calculate the total length of each of the two sketch segments. If we want to use $N+1$ points to define the central line, we divide each of the two total lengths by N to obtain $N+1$ points on each of the two sketch segments. That is to say, $\mathbf{P}_{1,n}$ ($n = 1, 2, 3, \ldots, N+1$) points on the left sketch segment, and $\mathbf{P}_{2,n}$ ($n = 1, 2, 3, \ldots, N+1$) points on the right sketch segment. The central line between the left sketch segment and the right sketch segment is generated from the points ($n = 1, 2, 3, \ldots, N+1$) determined by the equation below

$$\mathbf{C}_n = \mathbf{P}_{1,n} + 0.5(\mathbf{P}_{2,n} - \mathbf{P}_{1,n})$$
$$(n = 1, 2, 3, ..., N + 1) \tag{1}$$

By connecting all the points $\mathbf{C}_n$ together, we obtain the central line between the left sketch segment and the right sketch segment. The middle line shown in Fig. 3b gives the central line of the two sketch segments.

Next, we place a 3D cylinder to align with the two sketch segments. Since the central line is not straight, if straight cylinders without bending are chosen as 3D primitives, at least two or more cylinders have to be used and the space between two disconnected cylinders must be blended together to create the 3D left leg model. Clearly, this method of creating the 3D leg model is not a good choice. In this paper, we propose another method. With this method, a straight cylinder is bent by the central line. That is to say, the central line of the bent cylinder is the same as the central line between the two sketch segments. The radius of the bent cylinder is determined by the following equation

$$r = \frac{1}{N+1} \sum_{n=1}^{N+1} \|\mathbf{P}_{2,n} - \mathbf{P}_{1,n}\| / 2 \tag{2}$$

For the leg shown in Fig. 3b, we obtain $r = 0.537$ from Eq. (2). With the obtained radius, a straight cylinder is bent and placed along the central line. The obtained bent cylinder is shown in Fig. 3c.

The radius of the bent cylinder determined above is not optimal. It can be improved by minimizing the total error between the two projected silhouette curves of the bent cylinder and the two sketch segments. Similar to the above treatment, after projecting the bent cylinder to the plane of the two sketch segments, we obtain two projected silhouette curves. Then we draw straight lines passing through the points $\mathbf{P}_{1,n}$, $\mathbf{C}_n$, and $\mathbf{P}_{2,n}$, and calculate the intersecting points between the straight line and the two projected silhouette curves to obtain the $N+1$ points $\mathbf{C}_{1,n}$ on the left projected silhouette curve and $N+1$ points $\mathbf{C}_{2,n}$ on the right projected silhouette curve. If the radius of the bent cylinder is changed from r to $\bar{r}$, the left projected silhouette curve will be changed from $\mathbf{C}_{1,n}$ to $\bar{\mathbf{C}}_{1,n}$, and the right projected silhouette curve will be changed from $\mathbf{C}_{2,n}$ to $\bar{\mathbf{C}}_{2,n}$ where $\bar{\mathbf{C}}_{1,n}$ and $\bar{\mathbf{C}}_{2,n}$ are determined by the equations below

$$\bar{\mathbf{C}}_{1,n} = \mathbf{C}_n - \frac{\bar{r}}{r}(\mathbf{C}_n - \mathbf{C}_{1,n})$$
$$\bar{\mathbf{C}}_{2,n} = \mathbf{C}_n + \frac{\bar{r}}{r}(\mathbf{C}_{2,n} - \mathbf{C}_n)$$
$$(n = 1, 2, \ldots, N + 1) \tag{3}$$

The optimal radius $\bar{r}$ of the bent cylinder can be obtained by minimizing the sum of the squared errors between two new projected silhouette curves and the two sketch segments, i. e.,

$$\min_{\bar{r}} \sum_{n=1}^{N+1} \|\bar{\mathbf{C}}_{1,n} - \mathbf{P}_{1,n}\| + \|\bar{\mathbf{C}}_{2,n} - \mathbf{P}_{2,n}\| \tag{4}$$

Introducing Eq.(3) into Eq. (4), Eq.(4) is changed into

$$\min_{\bar{r}} \sum_{n=1}^{N+1} \|\mathbf{C}_n - \frac{\bar{r}}{r}(\mathbf{C}_n - \mathbf{C}_{1,n}) - \mathbf{P}_{1,n}\| + \|\mathbf{C}_n + \frac{\bar{r}}{r}(\mathbf{C}_{2,n} - \mathbf{C}_n) - \mathbf{P}_{2,n}\| \qquad (5)$$

Solving the minimization Eq. (5), we obtain the optimal radius $r = 0.574$. Using the value, we generate a new bent cylinder shown in Fig. 3d whose radius makes the two projected silhouette curves better approximate the two sketch segments. Figure 3e gives a comparison of the cylinders whose radii are determined by Eqs. (2) and (5), respectively.

5 Fitting 3D Primitives to 2D Sketches

The last step of our proposed approach is to develop an ordinary differential equation-based and sketch-guided deformation algorithm and use the algorithm to deform the placed cylinders so that the projected silhouette curves of the deformed cylinders exactly fit the two sketch segments of the decomposed parts. In what follows, we will elaborate how to develop such an ODE-based and sketch-guided deformation algorithm from a simplified version of the Euler-Lagrange PDE (Partial Differential Equation), which is widely used in physically-based surface deformations and briefly reviewed below.

As discussed in [2], the main requirement for physically-based surface deformations is an elastic energy which considers the local stretching and bending of two-manifold surfaces called thin-shells. When a surface $\mathbf{S} \subset \mathbf{R}^3$, parameterized by a function $\mathbf{P}(u, v) : \Omega \subset \mathbf{R}^2 \mapsto \mathbf{S} \subset \mathbf{R}^3$, is deformed to a new shape $\mathbf{S}'$ through adding a displacement vector $\mathbf{d}(u, v)$ to each point $\mathbf{P}(u, v)$, the change of the first and second fundamental $\mathbf{I}(u, v), \mathbf{II}(u, v) \in \mathbf{R}^{2 \times 2}$ forms in differential geometry [6] yields a measure of stretching and bending, as described in [26]:

$$\mathbf{E}_{shell}(\mathbf{S}') = \int_{\Omega} k_s \|\mathbf{I}' - \mathbf{I}\|_F^2 + k_b \|\mathbf{II}' - \mathbf{II}\|_F^2 \qquad (6)$$

where $\mathbf{I}'$ and $\mathbf{II}'$ are the first and second fundamental forms of the surface $\mathbf{S}'$, $\|.\|$ indicates a (weighted) Frobenius norm, and the stiffness parameters k_s and k_b are used to control the resistance to stretching and bending.

Generating a new deformed surface requires the minimization of the above Eq. (6), which is non-linear and computationally expensive for interactive applications. In order to avoid the nonlinear minimization, the change of the first and second fundamental forms is replaced by the first and second order partial derivatives of the displacement function $\mathbf{d}(u, v)$ [3,27], i. e.,

$$\tilde{\mathbf{E}}_{shell}(\mathbf{d}) = \int_{\Omega} k_s(\|\mathbf{d}_u\|^2 + \|\mathbf{d}_v\|^2) + k_b(\|\mathbf{d}_{uu}\|^2 + 2\|\mathbf{d}_{uv}\|^2 + \|\mathbf{d}_{vv}\|^2) \qquad (7)$$

where $\mathbf{d}_u = \frac{\partial}{\partial u}\mathbf{d}$, $\mathbf{d}_{uv} = \frac{\partial}{\partial u \partial v}\mathbf{d}$ and $\mathbf{d}_{uu} = \frac{\partial}{\partial u^2}\mathbf{d}$.

The minimization of the above equation can be obtained by applying variational calculus, which leads to the following Euler-Lagrange PDE:

$$- k_s \triangle \mathbf{d} + k_b \triangle^2 \mathbf{d} = 0, \tag{8}$$

In the equation, $\triangle$ and $\triangle^2$ are the Laplacian and the bi-Laplacian operators, respectively, which are defined below.

$$\triangle \mathbf{d} = div \nabla \mathbf{d} = \mathbf{d}_{uu} + \mathbf{d}_{vv},$$
$$\triangle^2 \mathbf{d} = \triangle(\triangle \mathbf{d}) = \mathbf{d}_{uuuu} + 2\mathbf{d}_{uuvv} + \mathbf{d}_{vvvv}. \tag{9}$$

Using the sketched 2D silhouette contours shown in Fig. 3b to change the shape of the cylinder can be transformed into the generation of a sweeping surface which passes through the two sketched 2D silhouette contours. The generator that creates the sweeping surface is a curve of the parametric variable u only, and the two silhouette contours are trajectories. If Eq. (8) is used to describe the generator, the parametric variable v in Eq. (8) drops, and we have $\mathbf{d}_{vv} = 0$ and $\mathbf{d}_{vvvv} = 0$. Substituting $\mathbf{d}_{vv} = 0$ and $\mathbf{d}_{vvvv} = 0$ into Eq. (9), we obtain the following simplified version of the Euler-Lagrange PDE, seen as (10), which is actually a vector-valued ODE.

$$k_b \frac{\partial^4 \mathbf{d}}{\partial u^4} - k_s \frac{\partial^2 \mathbf{d}}{\partial u^2} = 0. \tag{10}$$

As pointed out in [4], the finite difference solution to ODEs is very efficient. We here investigate such a numerical solution to Eq. (10). For a typical node i, the central finite difference approximations of the second and fourth order derivatives can be written as:

$$\frac{\partial^2 \mathbf{d}}{\partial u^2}|_i = \frac{1}{\triangle u^2}(\mathbf{d}_{i+1} - 2\mathbf{d}_i + \mathbf{d}_{i-1}),$$

$$\frac{\partial^4 \mathbf{d}}{\partial u^4}|_i = \frac{1}{\triangle u^4}[6\mathbf{d}_i - 4(\mathbf{d}_{i-1} + \mathbf{d}_{i+1}) + \mathbf{d}_{i-2} + \mathbf{d}_{i+2}]. \tag{11}$$

Introducing Eq. (11) into Eq. (10), the following finite difference equation at a representative node i can be written as:

$$(6k_b + 2k_s h^2)\mathbf{d}_i + k_b\mathbf{d}_{i-2} + k_b\mathbf{d}_{i+2} - (4k_b + k_s h^2)\mathbf{d}_{i-1} - (4k_b + k_s h^2)\mathbf{d}_{i+1} = 0. \tag{12}$$

For organic models, the 3D shape defined by two silhouette contours is closed in the parametric direction u as indicated in Fig. 4b. Therefore, we can extract some closed curves each of which passes through the two corresponding points on the two silhouette contours. Taking the silhouette contours in Fig. 4b as an example, we find two corresponding points $\mathbf{C}_{13}$ and $\mathbf{C}_{23}$ on the original silhouette contours $\mathbf{c}_1$ and $\mathbf{c}_2$, and two corresponding points $\mathbf{C}'_{13}$ and $\mathbf{C}'_{23}$ on the deformed silhouette contours $\mathbf{c}'_1$ and $\mathbf{c}'_1$ as shown in Fig. 4a. Then, we extract a closed curve $\mathbf{c}(u)$ passing through the two corresponding points $\mathbf{C}_{13}$ and $\mathbf{C}_{23}$ from

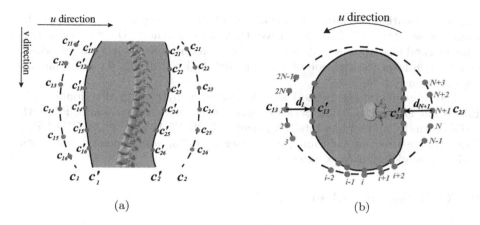

Fig. 4. Finite difference nodes for local shape manipulation from sketches in side and top view planes, respectively

the 3D model in Fig. 4a and depict it as a dashed curve in Fig. 4b. Assuming that the deformed shape of the closed curve $\mathbf{c}(u)$ is $\mathbf{c}'(u)$, the displacement difference between the original closed curve and the deformed closed curve is $\mathbf{d}(u) = \mathbf{c}'(u) - \mathbf{c}(u)$.

In order to use the finite difference method to find the displacement difference $\mathbf{d}(u)$, we uniformly divide the closed curve into $2N$ equal intervals as indicated in Fig. 4b. With the displacement difference at node 0 and node N already known, i. e. $\mathbf{d}_0 = \mathbf{C}'_{13} - \mathbf{C}_{13}$ and $\mathbf{d}_N = \mathbf{C}'_{23} - \mathbf{C}_{23}$, we can form a $2N$ linear algebra equations derived from (12) for each of these nodes' displacement. Solving the equations and adding all the displacement differences to the original curve $\mathbf{c}(u)$, we can then obtain the deformed curve $\mathbf{c}'(u)$, and depict it as a solid curve in

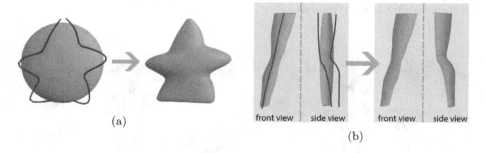

Fig. 5. ODE-driven Sketch-based deformations: (a)the deformation process of an organic shape represented by an ellipsoid and its 2D silhouette contour, and the deformed shape of the ellipsoid, (b) a leg that has been deformed by single-view sketch strokes before, now is further deformed in accordance with free form red-colored curves (Color figure online)

Fig. 4b. Repeating the above operations for all other points on the two silhouette contours, we obtain all deformed curves that describe a new 3D deformed shape. This method also applies to deformations responding to free form curves, which can be seen from the creation of a star model depicted in Fig. 5a and a human leg shown in Fig. 5b.

The aforementioned method was developed in python on the Houdini FX Education Edition 16.5.323 package, and ran on a dual boot Linux PC with 23 GB memory and 64 bits Intel(R) Xeon(R) CPU E5-1650 0 @ 3.20 GHz CPU. The average time for deforming a cylinder is 0.17 s, which ensures a smooth real-time modeling user experience.

6   Conclusion and Future Work

By extending the work described in [19], a sketch-based modeling approach has been developed in this paper. It creates base 3D models easily and quickly by first generating silhouette contours of 3D objects, decomposing the generated silhouette contours into parts to obtain the sketch segments of the parts, determining the central lines of parts, automatically placing ellipsoids or straight cylinders by bending the cylinders along the central lines to best align with the sketch segments, scaling the placed ellipsoids to match the sketch segments, minimizing the sum of squired errors between the projected silhouette curves of the placed cylinders and the sketch segments to obtain the optimal radius of the placed cylinders and make the placed cylinder best fit the sketch segments. Furthermore, an ordinary differential equation (ODE)-based deformation algorithm has been developed to deform the placed cylinders so that the projected silhouette curves of the deformed cylinders are exactly the same as the sketch segments.

With our proposed approach, different 3D base models were created from 2D sketches and depicted in Fig. 6. These created base 3D models indicate that our proposed approach achieves the advantages of: 1) easiness for beginners to use, 2) avoiding heavy manual operations, and 3) high efficiency in creating sketch-based 3D organic models.

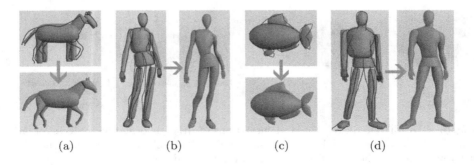

(a)   (b)   (c)   (d)

Fig. 6. Examples of the primitive deformation method

Up to now, various sketch-based 3D modeling approaches can only create rough 3D models. How to develop new sketch-based modeling approaches to create detailed and realistic 3D models still has a long way to go. The approach proposed in this paper tackles how to create rough 3D base models easily and quickly. In our following work, we will further investigate various sketch-guided and ODE-based deformation methods to obtain detailed deformations and create local 3D shapes for more realistic sketch-based 3D modeling.

Acknowledgements. This research is supported by the PDE-GIR project which has received funding from the European Unions Horizon 2020 research and innovation programme under the Marie Skodowska-Curie grant agreement No. 778035, and Innovate UK (Knowledge Transfer Partnerships Ref: KTP010860).

References

1. Botsch, M., Kobbelt, L.: An intuitive framework for real-time freeform modeling. ACM Trans. Graph. **23**(3), 630–634 (2004)
2. Botsch, M., Sorkine, O.: On linear variational surface deformation methods. IEEE Trans. Vis. Comput. Graph. **14**(1), 213–230 (2008)
3. Celniker, G., Gossard, D.: Deformable curve and surface finite-elements for free-form shape design. ACM SIGGRAPH Comput. Graph. **25**(4), 257–266 (1991)
4. Chaudhry, E., You, L.H., Jin, X., Yang, X.S., Zhang, J.J.: Shape modeling for animated characters using ordinary differential equations. Comput. Graph. **37**(6), 638–644 (2013)
5. Chen, T., Zhu, Z., Shamir, A., Hu, S.M., Cohen-Or, D.: 3-sweep: extracting editable objects from a single photo. ACM Trans. Graph. **32**(6), 195 (2013)
6. Do Carmo, M.P., Fischer, G., Pinkall, U., Reckziegel, H.: Differential geometry. In: Mathematical Models, pp. 155–180. Springer (2017)
7. Facebook: Quill (2019). https://quill.fb.com/
8. Gingold, Y., Igarashi, T., Zorin, D.: Structured annotations for 2D-to-3D modeling. ACM Trans. Graph. **28**(5), 148 (2009)
9. Google Inc.: Tilt brush (2017). https://www.tiltbrush.com/
10. Gravity Sketch: Gravity sketch (2017). https://www.gravitysketch.com/
11. Hoffman, D.D., Singh, M.: Salience of visual parts. Cognition **63**(1), 29–78 (1997)
12. Igarashi, T., Moscovich, T., Hughes, J.F.: As-rigid-as-possible shape manipulation. ACM Trans. Graph. **24**(3), 1134–1141 (2005)
13. Karpenko, O.A., Hughes, J.F.: Smoothsketch: 3D free-form shapes from complex sketches. ACM Trans. Graph. **25**(3), 589–598 (2006)
14. Kazmi, I.K., You, L.H., Yang, X.S., Jin, X., Zhang, J.J.: Efficient sketch-based creation of detailed character models through data-driven mesh deformations. Comput. Animation Virtual Worlds **26**(3–4), 469–481 (2015)
15. Kazmi, I.K., You, L.H., Zhang, J.J.: A survey of sketch based modeling systems. In: The 11th International Conference on Computer Graphics, Imaging and Visualization (CGIV), pp. 27–36. IEEE (2014)
16. Kazmi, I.K., You, L.H., Zhang, J.J.: A hybrid approach for character modeling using geometric primitives and shape-from-shading algorithm. J. Comput. Des. Eng. **3**(2), 121–131 (2016)

17. Kraevoy, V., Sheffer, A., van de Panne, M.: Modeling from contour drawings. In: Proceedings of the 6th Eurographics Symposium on Sketch-Based interfaces and Modeling, pp. 37–44. ACM (2009)
18. Li, C., Pan, H., Liu, Y., Tong, X., Sheffer, A., Wang, W.: Bendsketch: modeling freeform surfaces through 2D sketching. ACM Trans. Graph. **36**(4), 125 (2017)
19. Li, O., et al.: ODE-driven sketch-based organic modelling. In: Gavrilova, M., Chang, J., Thalmann, N., Hitzer, E., Ishikawa, H. (eds.) Advances in Computer Graphics. CGI 2019. Lecture Notes in Computer Science, vol. 11542, pp. 453–460 (2019)
20. Mozilla: Mozilla a-painter (2019). https://aframe.io/a-painter/
21. Nealen, A., Igarashi, T., Sorkine, O., Alexa, M.: Fibermesh: designing freeform surfaces with 3D curves. ACM Trans. Graph. **26**(3), 41 (2007)
22. Nealen, A., Sorkine, O., Alexa, M., Cohen-Or, D.: A sketch-based interface for detail-preserving mesh editing. In: ACM SIGGRAPH 2007 Papers - International Conference on Computer Graphics and Interactive Techniques, pp. 42–47. ACM (2007)
23. Oculus: Oculus medium (2019). https://www.oculus.com/medium/
24. Olsen, L., Samavati, F.F., Sousa, M.C., Jorge, J.A.: Sketch-based modeling: a survey. Comput. Graph. **33**(1), 85–103 (2009)
25. Shtof, A., Agathos, A., Gingold, Y., Shamir, A., Cohen-Or, D.: Geosemantic snapping for sketch-based modeling. Comput. Graph. Forum **32**(2), 245–253 (2013)
26. Terzopoulos, D., Platt, J., Barr, A., Fleischer, K.: Elastically deformable models. ACM SIGGRAPH Comput. Graph. **21**(4), 205–214 (1987)
27. Welch, W., Witkin, A.: Variational surface modeling. ACM SIGGRAPH Comput. Graph. **26**(2), 157–166 (1992)
28. Wuhrer, S., Shu, C.: Shape from suggestive contours using 3D priors. In: Proceedings of the Ninth Conference on Computer and Robot Vision, pp. 236–243 (2012). https://doi.org/10.1109/CRV.2012.38
29. Xu, M., Li, M., Xu, W., Deng, Z., Yang, Y., Zhou, K.: Interactive mechanism modeling from multi-view images. ACM Trans. Graph. **35**(6), 236 (2016)
30. You, L.H., Ugail, H., Tang, B.P., Jin, X., You, X.Y., Zhang, J.J.: Blending using ODE swept surfaces with shape control and C^1 continuity. Visual Comput. **30**(6–8), 625–636 (2014)
31. You, L.H., Yang, X.S., Pachulski, M., Zhang, J.J.: Boundary constrained swept surfaces for modelling and animation. Comput. Graph. Forum **26**(3), 313–322 (2007)
32. You, L.H., Yang, X.S., You, X.Y., Jin, X., Zhang, J.J.: Shape manipulation using physically based wire deformations. Comput. Animation Virtual Worlds **21**(3–4), 297–309 (2010)
33. Zimmermann, J., Nealen, A., Alexa, M.: Sketch-based interfaces: sketching contours. Comput. Graph. **32**, 486–499 (2008). https://doi.org/10.1016/j.cag.2008.05.006

Tree Species Modelling for Digital Twin Cities

Like Gobeawan[1(✉)], Daniel J. Wise[1], Sum Thai Wong[1], Alex T. K. Yee[2],
Chi Wan Lim[1], and Yi Su[1]

[1] Institute of High Performance Computing, A*STAR, Singapore, Singapore
`gobeawanl@ihpc.a-star.edu.sg`
[2] National Parks Board, Singapore, Singapore

Abstract. Creating vegetation contents for a digital twin city entails
generating dynamic 3D plant models in a large scale to represent the
actual vegetation in the city. To enable high-fidelity environmental simulations and analysis applications, we model individual trees at a species
level of detail. The 3D models are generated procedurally based on their
botanical species profiles within the constraints of measurements and
growth spaces derived from laser-scanned point cloud data. Users can
conveniently define the known profile of a species by using a species profile template that we formulated based on species growth processes and
patterns. Based on the given species profile and solving for the unknowns
within the growth space constraints, a species model will be grown
through iterations of our formulated growth rules. We show that this
methodology produces structurally-representative species models with
respect to their actual physical and species characteristics.

Keywords: Tree modelling · Tree species architecture · Species
growth · Procedural modelling · Digital twins · Optimisation

1 Introduction

The development of digital twin cities, as the digital representation of actual
cities, is on the rise around the world [33] for their role and importance to
decision makers in monitoring and planning various urban aspects. Based on
remote sensing data, a digital twin city are populated with representative urban
objects of a real city, including the species-diverse, dynamically-growing trees. To
represent millions of individual trees for a typical green city like Singapore [29],
static generic 3D trees are commonly adopted [20], in contrast to urban buildings
which most existing works have been focused on and are modeled at various
resolutions up to an interior level of detail [1]. Due to their challenges, trees
are often modelled regardless of their species characteristics such as their actual
growth and branching structure. On the other hand, a more detailed tree species
representation in the digital twin cities will enable more useful simulations and
analysis for city and environment planning, forestation, food and agriculture
strategies, and so on.

M. L. Gavrilova and C. J. K. Tan (Eds.): Trans. on Comput. Sci. XXXVIII, LNCS 12620, pp. 17–35, 2021.
https://doi.org/10.1007/978-3-662-63170-6_2

Modelling tree species for digital twin cities poses challenges of scalability, dynamism, representative structure, and extendability to all species. We aim to address these challenges by modelling the actual physical and biological appearance and growth of the species. Specifically, given tree point clouds from the remote sensing data and the species information, we propose to automatically generate dynamic 3D species models with respect to individual species-specific architecture and morphology.

In essence, our work presents the following contributions.

1. An automatic mechanism to generate digital twins of trees at a species level based on biological and physical species traits that fit the remote sensing data constraints
2. A species profile template that enables user-convenient species modelling and is extendable to new species
3. Various species profiles and species models for ten tropical tree species
4. Biologically representative growth rules for all varieties of tree species
5. Optimisation strategy to fit and grow tree species models within the point cloud data constraints

A preliminary work-in-progress version of this species modelling work has been reported in [10], along with a series of our other related works [8,9,19–21] to model trees at various levels of details for Virtual Singapore. Further works on the species profile template, the species profiles for various species, the growth rules, and optimisation results are covered here in Sects. 3 and 4.

2   Literature Review

Many approaches of 3D tree modelling have been proposed for different purposes. Interactive techniques such as [23,25,38] require manual intervention to generate realistic individual models, hence they are generally not suitable for large-scale modelling. Image-based techniques [16,32] and point cloud reconstruction [14, 24,39] are generally scalable albeit producing static non-species representative models. On the other hand, procedural approaches produce a vast variety of trees with believable branching structure automatically based on a relatively small set of rules. They are potentially suitable for large scale modelling in virtual cities. However, most procedural approaches are stochastic in nature [17,26,38,40], making stochastic procedural approaches unsuitable to control the outcome to represent actual trees in digital twin cities.

Non-stochastic procedural models are generally suitable for representing actual individual trees in a large scale setting of a virtual city. The works of [8,20] generate static, simplified, low resolution models of real trees for a virtual city at a level of detail that represent crown shapes and tree heights. Boudon et al. [5] as well as Bernard and McQuillan [3] constructed tree skeletons which match those of the point clouds of trees with sparse foliage. Similarly, Stava et al. [35] formulated a generic, not species-specific, parameterised model to generate a similar tree skeleton through semi-stochastic Monte Carlo Markov

Chain optimisation [18], which also worked well for trees with sparse foliage. The resulting models are realistic and dynamically growing. In addition, most FSPM (functional structural plant modelling)-based works [12,30,34,36,37,41] produce dynamic, accurate individual or collective tree models in response to environments, although they tend to be valid only for certain species or generic trees and often require extensive biomass data, unrelated to remote sensing data. Our proposed methodology [10] is a FSPM-based species modelling to fit within the given point cloud growth space. In our case, the FSPM-based growth rules focus on species architecture facts which users can readily provide based on their knowledge and field surveys. The resulting models are optimised to match the growth space and known measurement values of point cloud data.

3 Methodology

To generate tree species models, we propose a procedural methodology comprising of three major components: species profile template, species growth rules, and parameter optimisation within data constraints (Fig. 1).

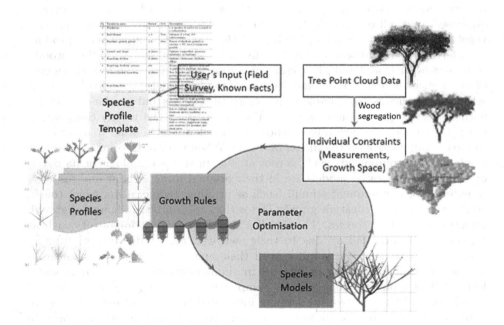

Fig. 1. Tree species modelling workflow

Firstly, we have formulated a species profile template, which contains relevant parameters that define species-specific growth processes and patterns. A user can fill in the template with known parameter values or ranges based on field surveys and known facts of a species, creating a species profile for each species. Given a species name, its corresponding species profile of known parameter values will be fed into our formulated growth rules to generate a species model after solving for the unknowns by a parameter optimisation.

The growth rules, an L-system [22,31]-based production rules, dictate the species life cycle from the seed until the maturity and the senescence. The parametric rules describe the hierarchical development of tree components such as root, trunk, branch, leaf, flower, fruit, and potentially their internals down to the cellular level. The rules were constructed specifically to match the species architecture and characteristics of a target species specified by the parameters in the species profile.

Unknown parameters in the growth rules are then solved by the parameter optimisation for each individual tree. The optimisation is constrained by individual tree measurements and growth spaces derived from remote sensing point cloud data [19,21] to grow a species model within a growth space, which is the space occupied by a tree point cloud. This results in a species model similar to the actual tree in terms of physical size, growth, branching pattern, and morphology.

In this work, we made a few assumptions in modelling tree species from remote sensing data. Firstly, trees are tropical plants, whose architecture is organised and well recognised [13]. Secondly, trees are mainly modelled in their vegetative/growth stage of life, leaving out the stages of germination, reproduction, fertilisation, and senescence. Thirdly, trees are represented by their woody components such as trunk, branches, and root, while non-woody components of trees such as leaf, flower, and fruit are loosely modeled (random) as their seasonal presence is not reliable to represent the species and can be established with respect to given simulation conditions at a time. The woody components of the tree models are generated without any randomness in order to allow users to have a full control on the model outcome. When data or information are not available, tree components such as root are assigned a placeholder for its development. The generated models are in their original form without any response to external or environment stimuli (such as pruning, temperature, wind, touch, etc.), except for the constant gravity factor that affects the fundamental growth direction of all tree species. This implies that the woody components such as branches will never fall off due to their own weights, wind, rain, or pruning. Trees will continue to grow taller until their apical buds abort or transform. When these models are actually used in simulations, external environmental forces will act on and change their components accordingly. For the sake of simplicity and user control, we assume individual tree components conform to a single growing process, uniform branching pattern, and similar response to external stimuli. Specifically, a tree model will have the same branching angles for all its branches and a single growth model of woody components throughout its lifespan. In nature, trees usually have some branching angle variations in response to the dynamic environment, or different growth models at different life stages. A more detailed, accurate model with various growth changes can be produced when the corresponding accurate data and knowledge are input into the model generator.

The following Sects. 3.1, 3.2, and 3.3 will describe each component of the species modelling methodology in detail.

3.1 Species Growth Rules

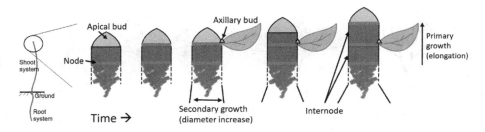

Fig. 2. Fundamental growth of plants

In general, trees go through stages of life cycle from germination, vegetative/growth, reproduction/maturity, pollination/fertilisation, and senescence. The life stages are captured in our formulated growth rules that govern the development of various tree components over time. In the focus of the growth rules is the growth stage, when trees undergo a primary growth (stem elongation) near apical buds and a secondary growth (trunk diameter increase) further away from the buds as illustrated in Fig. 2. The rules incorporate growth aspects of the tree architecture (Fig. 3) as described in [2]. When these aspects are combined differently, they will form 23 tree architecture models [13] covering all (tropical) tree species in nature.

We adopted L-system [22, 31] in implementing the growth rules. Our species growth rules define transformation (from the left hand side to the right hand side of a rule) among basic components and process modules as outlined in Table 1. Each module encapsulates its own growth mechanism involving a set of parameters, whose known values are given by species profiles and unknown values are solved by a parameter optimisation. At each iteration or time step, when a condition is fulfilled, suitable rules will be activated. Given a species profile with a specific parameter configuration, the growth rules will produce a growing species model over time.

3.2 Species Profiles

The growth rules take in values of parameters from a species profile to generate a species model. The species profile parameters can be grouped into five categories as shown in Table 2: growth process (primary and secondary growths), branching process, phyllotaxis, response to external stimuli, and constraint parameters.

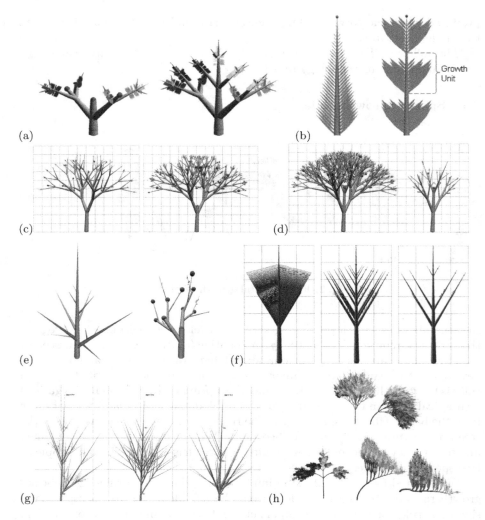

Fig. 3. Tree structural architecture patterns: (a) determinate/indeterminate growth, (b) continuous/rhythmic growth, (c) terminal/lateral branching, (d) immediate/delayed branching, (e) monopodial/sympodial branching, (f) continuous/rhythmic/diffuse branching, (g) acrotonic/mesotonic/basitonic branching, (h) orthotropy/plagiotropy. Red spheres at the tips of branches indicate buds that have aborted or transformed to terminal organs such as flowers. (Color figure online)

Constraint parameters are individual tree measurement parameters derived from point cloud data or field measurements which constrain a generated model to be similar to actual trees (Fig. 4). Examples of constraint parameters are

Table 1. Growth rules

Life cycle phase	Rule
Germination	Seed → Root Shoot
	Root → Root
	Shoot → Bud
	Bud → Bud
Primary growth	Bud → Node Internode Bud
	Node → Leaf [Bud] Node
Secondary growth	Internode → Internode
	Node → Node
Terminal branching	Bud → [Bud] [Bud]
Leaf growth	Leaf → Leaf
	Leaf →
Maturity	Bud → Flower
	Flower → Flower
	Flower →
Fertilisation	Flower → Seed Fruit
	Fruit → Fruit
	Fruit →
Senescence	Bud →

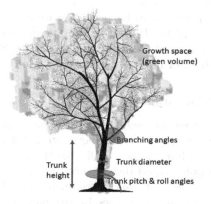

Fig. 4. Examples of constraint parameters

trunk height, trunk diameter, branching angle, and growth space. The trunk height parameter restricts the first tree branching to start at the trunk height, while trunk diameter and branching angle parameters restrict the age and spread of the tree crown to fit within the growth space. A collection of constraints of many individual trees of a species will form a species statistics to predict the unknowns.

On the other hand, the remaining profile parameters define the natural tree growth of a species. Their values can be derived from knowledge, statistics, biological facts, field measurements, or observation. Examples of such parameters are bud lifespan, rhythmic growth pattern, diameter growth, and branch arrangement.

Diameter growth, in particular, varies greatly among different species and individuals. As such, we use the relative growth rate (RGR) to represent the diameter growth rate across the tree body, taking one of the growth models commonly used by ecologists [7]: linear, exponential, power law, monomolecular, three-parameter logistic, four-parameter logistic, and Gompertz models. In our field experiment work [20], we measure the trunk diameter at 1.3-meter breast-height in field measurement to represent the tree biomass in calculating RGR. We then apply the growth model to estimate the diameters of all branches at any point of time. When field measurement data are not available, users can adopt one of the growth models and let the system estimate the parameters.

In fact, for all parameters in the species profile template, users only need to provide known range values (including the growth space derived from point cloud data, if any), and the system will solve for the unknowns by an optimisation in order to generate the target species models within reasonable ranges based on existing statistics and facts. In this work, we solve for the constraint parameters in Table 3 while keeping the values of other parameters fixed according to the user-specified tree species profile and the growth space.

Users may model a new species by providing its known species information into the species profile template (Table 2). Examples of completed species profiles for ten Singapore major species are listed in Table 3. Species profiles will adopt a default value when the user does not specify an input to a profile parameter.

3.3 Parameter Optimisation

Along with known measurements and the growth space constraints derived from point cloud data, there generally exist parameters with uncertain, unknown values such as age, branching angles, and growth rate. Such parameters are unattainable from the point cloud data either due to data noise or absence of spatial-temporal point cloud data and periodical field observations. We solve for uncertain, unknown parameter values by the parameter optimisation using a genetic algorithm [28] as illustrated in Fig. 5.

An initial population of 1000 possible parameter configuration solutions is generated using Latin hypercube sampling [27] over ranges of possible values specified in the species profile. The cost function is evaluated for each solution and the solution-cost entry is appended to a database. After testing the initial population, the algorithm iteratively selects a number of best solutions (of lowest cost) as well as a number of random solutions from the database, collectively as the parent solutions. They are combined in pairs, and their parameter values are

Table 2. Species profile template

No.	Parameter name	Format	Unit	Description
A	**Growth process**			
A1	**Primary growth**			
1	Bud lifespan	≥ 0	year	Lifespan of a bud, 0 if indeterminate
2	Rhythmic growth period	≥ 0	year	Period of rhythmic growth to produce 1 GU, zero if continuous growth
3	Growth unit shape	a choice	–	Unspecified (U), acrotonic (A), mesotonic (M), or basitonic (B)
A2	**Secondary growth**			
4	Diameter growth model	a choice	–	Relative growth rate (RGR): linear (L), exponential (E), power law (P), monomolecular (M), 3-param logistic (3PL), 4-param logistic (4PL), Gombertz (G)
5	Initial diameter	> 0	meter	Non-zero minimum branch diameter
6	Diameter growth variables	r, β, K, L	ranges, t: day, M: cm, ln(M)	Variables for RGR equation with respect to growth model
B	**Branching process**			
7	Branching rhythm	a choice	–	Continuous (C), rhythmic (R), or diffuse (D)
8	Branching rhythmic pattern	xxx	–	Binary rhythmic pattern of 0s and 1s for rhythmic branching
9	Terminal/lateral branching	a choice	–	New branches are formed by lateral buds (L) or apical split (T)
10	Branching delay	≥ 0	year	New branches grow out immediately (0) or after some delay
11	Monopodial/sympodial branching	a choice	–	Apical stem remains dominant with emergence of lateral branches (M) or stops growing with emergence of dominant lateral branches (S)
12	Number of apices	≥ 1	–	Sole or multiple number of dominant apices (scaffolds) at a time

(continued)

Table 2. (*continued*)

No.	Parameter name	Format	Unit	Description
C	**Phyllotaxis**			
13	Branch arrangement	a choice	–	Alternate (1), opposite (2), or whorled (n)
14	Divergence angle	$0 \leq \theta \leq 180$	degree	Angle between two branches from subsequent rows
D	**Response to external stimuli**			
15	Axis morphology	a matrix	–	Orientation of proximal and distal axes due to various tropisms; options for gravitropism: orthotropic (O) or plagiotropic (P)
16	Pruning diameter	≥ 0	meter	To discard branch part with diameter of or below pruning diameter
E	**Constraint parameters**			
17	Growth space	a volume	–	Physical space occupied by the tree
18	Age	a range, ≥ 1	year	Typical age for the species
19	Trunk pitch angle	a range, $0 \leq \theta \leq 180$	degree	Angle between up vector and trunk direction at bottom
20	Trunk roll angle	a range, $0 \leq \theta < 360$	degree	Counter-clockwise (CCW) angle from North to trunk on ground
21	Trunk height	a range, ≥ 0	meter	Shortest distance from ground/flare to lowest crown or first branching point
22	Number of first branches	a range, ≥ 0	–	At first branching point from bottom
23	Branching pitch angle	a range, $0 \leq \theta \leq 180$	degree	From parent's head down to start of branch
24	Branching roll angle	a range, $0 \leq \theta < 360$	degree	Rotate CCW around parent's head
25	Diameter growth rate	a range, $0 \leq g \leq 1$	–	Normalised growth rate; first three decimals are mapped to diameter growth variables
26	Number of new nodes	a range, > 0	/year	Number of nodes a bud produces in a year
27	Internode length	a range, ≥ 0	meter	Length of a segment between two consecutive nodes

Table 3. Examples of species profiles

No.	Parameter name	A. alexandrae	H. odorata	K. senegalensis	P. pterocarpum	S. grande	S. macrophylla	S. myrtifolium	S. parviflora	S. saman	T. rosea
1	Bud lifespan	0	30	5	10	30	30	8	10	5	5
2	Rhythmic growth period	0	2	1	2	0	5	0	2	0	0
3	Growth unit shape	u	U	U	U	U	U	U	U	U	U
4	Diameter growth model	M	L	L	L	P	L	M	E	L	E
5	Initial diameter	0.05	0.01	0.01	0.05	0.1	0.08	0.01	0.01	0.02	0.01
6	Diameter growth variables	r, K	r	r	r	r, β	r	r, K	r	r	r
7	Branching rhythm	C	R	R	R	R	R	R	R	R	R
8	Branching rhythmic pattern	0	$0^{26}10^3 10^{20}1$	$0^{44}1$	$0^{35}10^{20}1$	$0^9 10^6 1$	$0^{28}10^8 1$	$0^{14}1$	$0^9 1$	$0^{43}1$	$0^{14}10^7 1$
9	Terminal/lateral branching	L	L	L	L	L	L	L	L	L	L
10	Branching delay	∞	3	1	2	2	2	1	1	1	1
11	Monopodial/sympodial	M	M	M	S	S	M	S	M	M	S
12	Number of apices	1	1	1	many	1	1	1	1	many	1
13	Branch arrangement	1	1	1	1	2	1	2	1	1	2
14	Divergence angle	137.5	137.5	137.5	137.5	45.0	137.5	90.0	180.0	137.5	90.0
15	Axis morphology	O, P	O, P	O	O, P	O	O	O	O, P	O	O
16	Pruning diameter	0	0	0	0	0	0	0	0	0	0

Table 4. Optimised parameters for various tree species samples

Parameters	A. alexandrae	H. odorata	K. senegalensis	P. pterocarpum	S. grande	S. macrophylla	S. myrtifolium	S. parviflora	S. saman	T. rosea
Age	10	32	12	48	25	39	14	18	48	21
Trunk pitch angle	0.1800	0.0000	12.9400	0.4400	0.0000	11.2300	7.2900	0.0000	6.8800	1.6400
Trunk roll angle	30.6000	119.3300	43.2400	217.7700	194.4400	64.9500	224.5900	229.9800	195.0700	0.0000
Trunk height	7.3600	2.3800	4.3400	1.6900	2.4200	3.5300	0.9500	3.7700	1.1300	2.2500
Number of 1st order branches	-	2	2	1	1	3	3	2	2	3
Branch pitch angle	8.4505	47.8612	30.0000	20.0000	20.3371	27.5231	30.5086	42.3422	31.7527	15.3773
Branch roll angle	110.0919	138.6670	167.1365	137.7591	46.5599	143.4024	91.3802	120.8121	311.9920	257.4362
Diameter growth rate	0.0465	0.1000	0.9000	0.1590	0.8400	0.2900	0.0017	0.0010	0.0010	0.4470
Number of new nodes/year/bud	10	24	145	14	8.0000	15	14	30	25	9.0000
Internode length	0.0962	0.0099	0.0262	0.0240	0.0852	0.0266	0.0364	0.0204	0.0128	0.0680

mutated (slightly modified) to form a successive generation of 4 child solutions. The successive generations are then evaluated for their costs, and the solution-cost entries are appended to the database. The algorithm terminates after a specified number of generations or when a solution is under a specified cost threshold, whichever is earlier.

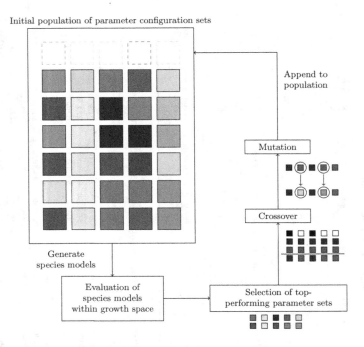

Fig. 5. Genetic algorithm workflow for optimising tree parameters

The cost function E for the optimisation is a linear weighted distances between a proposed solution and the target tree constraints. The target constraints are a combination of the macroscopic shape of the target tree (i.e., crown shape and dimension) and the microscopic structure of the target tree (i.e., measurements, growth space). Specifically, the growth space is a set of voxels within a uniform grid in the 3D space. Each voxel represent a space containing at least a threshold number of points of the point cloud.

$$E = \sum_{i=1}^{5} w_i E_i \tag{1}$$

$$E_1 = 1 - \frac{|C_L \cap C_G|}{|C_G|} \tag{2}$$

$$E_2 = \frac{|C_G \backslash C_L| + |C_L \backslash C_G|}{|C_G|} \tag{3}$$

$$E_3 = \frac{|\mathbf{B}_{L_{max}} - \mathbf{B}_{G_{max}}| + |\mathbf{B}_{L_{min}} - \mathbf{B}_{G_{min}}|}{|\mathbf{B}_{G_{max}} - \mathbf{B}_{G_{min}}|} \tag{4}$$

$$E_4 = \frac{|g_L - g_G|}{max(g_L, g_G)} \tag{5}$$

$$E_5 = \sum_i \frac{|a_{i_L} - a_{i_G}| + |b_{i_L} - b_{i_G}|}{a_{i_G} + b_{i_G}} \tag{6}$$

w_i is the weight of the error cost component E_i, which captures a certain geometrical difference between the solution tree and the growth space.

E_1 measures the empty space unfilled by the solution tree within the growth space. C_L is the set of voxels occupied by the tree and C_G is the set of voxels in the growth space.

E_2 measures the extra space occupied by the solution tree outside the growth space.

E_3 measures the difference between the bounding boxes of the tree and the growth space. $\mathbf{B}_{L_{min}}$ and $\mathbf{B}_{L_{max}}$ are two diagonally-opposite corner points of the boundary box of the tree, while $\mathbf{B}_{G_{min}}$ and $\mathbf{B}_{G_{max}}$ are those of the growth space, correspondingly.

E_4 measures the difference in the trunk girths of the tree and the growth space at a certain height. g_L and g_G are the trunk girths of the tree and the growth space, respectively. In particular, E_4 drives the setting of the appropriate diameter growth and age of the tree.

E_5 measures the difference in the overall shape of the crown by comparing the radii of the smallest horizontal ellipses which bound the crowns of the tree and the growth space at various heights of the crown. a_{i_L}, b_{i_L}, a_{i_G}, and b_{i_G} are the radii a and b of the ellipses at the height i of the crown of the tree L and the growth space G. E_5 tends to set branching angle parameters.

The parameter optimisation results in a set of constraint parameter configuration, which are used, along with other known parameter values, by the growth rules to generate a species model that is similar to the actual tree in the field.

4   Results

Growth rules were implemented in Python using L-Py [5] and tested to generate 10 tropical tree species models with average parameter values obtained from actual field measurements. The resulting species tree models are shown in Fig. 6.

Based on our processed point cloud data [8], the species profile, and the growth space of actual tree inputs, we also experimented to generate species models that fit within given growth spaces by optimising ten constraint parameters, within a limited runtime (12 h per task thread with 100 GB memory on an i7-7800X CPU) or until the error E stabilises, whichever is earlier. The

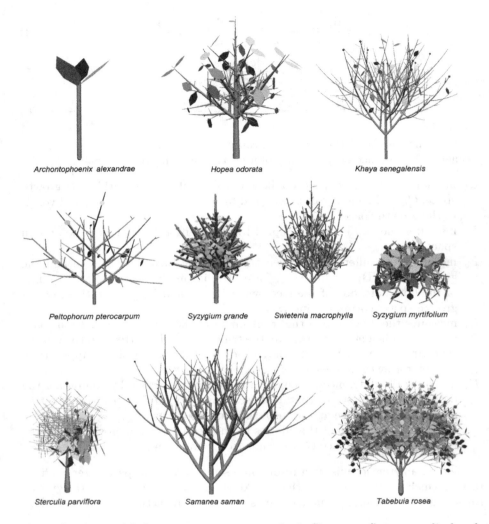

Fig. 6. Species models for ten common tree species in Singapore (leaves are displayed only for decoration): (a) *Archontophoenix alexandrae*, (b) *Hopea odorata*, (c) *Khaya senegalensis*, (d) *Peltophorum pterocarpum*, (e) *Syzygium grande*, (f) *Swietenia macrophylla*, (g) *Syzygium myrtifolium*, (h) *Sterculia parviflora*, (i) *Samanea saman*, and (j) *Tabebuia rosea*.

optimised parameter results are shown in Table 4, and their corresponding tree models within given growth spaces (voxel spacing of 50cm, minimum of 5 points per voxel) are shown in Fig. 7.

The results show that relatively similar, realistically structured species models can be automatically generated given the growth spaces of actual trees of known species without any measurement information. The branching topology

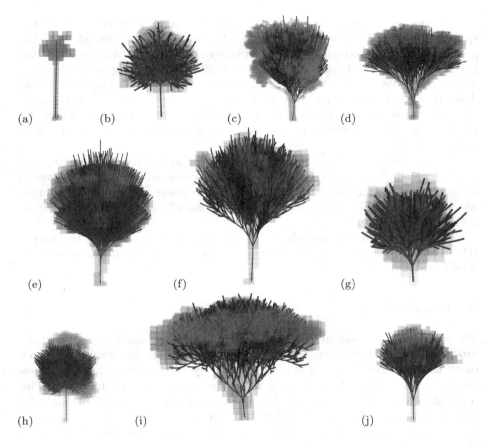

Fig. 7. Optimised species models and superimposed with growth spaces of actual trees: (a) *A. alexandrae* ($E_{1,2} = 83.69\%$), (b) *H. odorata* ($E_{1,2} = 24.64\%$), (c) *K. senegalensis* ($E_{1,2} = 95.26\%$), (d) *P. pterocarpum* ($E_{1,2} = 62.30\%$), (e) *S. grande* ($E_{1,2} = 30.81\%$), (f) *S. macrophylla* ($E_{1,2} = 43.49\%$), (g) *S. myrtifolium* ($E_{1,2} = 33.08\%$), (h) *S. parviflora* ($E_{1,2} = 49.86\%$), (i) *S. saman* ($E_{1,2} = 65.35\%$), (j) *T. rosea* ($E_{1,2} = 28.00\%$)

of the generated models does not necessarily match that of actual trees, but the branching patterns are those typical of their species. In comparison, other tree modeling works such as [3,5,35] reconstruct tree models from point cloud data without considering growth and branching structure specific to the tree species. The point cloud input data inherently contain noises, incomplete missing points, or fuzzy connections among the points, especially at the tree tips with fine branches. As the result, the reconstructed branch topologies are usually accurate at low branching orders (trunk or big branches), but may not reflect actual species' branching pattern or other botanical species features for fine, small branches.

Our results also show noticeable variably-high minimised error for all ten species (24.64%–95.26%) which seem to reflect a poor, loose model fit to the growth space of actual data. They are mainly accounted by the facts that the generated species models do not include non-wooden tree components (such as leaves, fruits, flowers, etc.) and do not consider trees' growth responses to the dynamic environment (such as human pruning, temperature, sunlight, wind, rain, etc.). Incorporating tree responses to the environment will potentially improve the fitting result, and this will form our future work as highlighted in Sect. 5 below.

We also showed that our models can represent actual trees with a constant growth model. Some species such as *Tabebuia rosea* undergo different modes of growth [4] throughout their life cycles, hence users need to specify multiple growth models in their species profiles to model the species growth more accurately.

The generated species models can be stored in multi-scale tree graph format [11] for subsequent post-processing such as mesh generation [19] in order to be used in simulations.

5   Conclusions

We presented an automated methodology to grow tree species models within growth spaces by our formulated growth rules and species profiles. The methodology can also model new, additional species by getting users to provide new species profiles with the help of a convenient species profile template. This methodology streamlines tree model generation from remote sensing data for digital twins and subsequently enables a better botanical representation of trees for urban modelling and simulation use cases such as wind-tunnel simulation [6,15].

This work can be further improved by incorporating the tree reiteration [2] and responses to external stimuli to further shape the tree models, as well as the natural fall of old branches due to mechanical factors (own weight, tree stability, etc.). For a better fit of species models within given growth spaces, the system should optimise for more parameters and multiple growth models instead of adhering to the fixed species profile.

Another future work in the pipeline is the full automation of species profiling without users' manual inputs by using machine learning techniques on the remote sensing data. This will streamline the entire framework from remote sensing data to modeling automatically without manual input of species information.

Acknowledgments. This work is supported by National Research Foundation Singapore, Virtual Singapore Award no. NRF2015VSG-AA3DCM001-034. Authors thank colleagues at IHPC (A*STAR), NParks, and GovTech for their valuable input and support.

References

1. Arroyo Ohori, K., Biljecki, F., Kumar, K., Ledoux, H., Stoter, J.: Modeling cities and landscapes in 3D with CityGML. In: Borrmann, A., König, M., Koch, C., Beetz, J. (eds.) Building Information Modeling, pp. 199–215. Springer, Cham (2018). https://doi.org/10.1007/978-3-319-92862-3_11
2. Barthélémy, D., Caraglio, Y.: Plant architecture: a dynamic, multilevel and comprehensive approach to plant form, structure and ontogeny. Ann. Bot. **99**, 375–407 (2007)
3. Bernard, J., McQuillan, I.: A fast and reliable hybrid approach for inferring L-systems. In: The 2018 Conference on Artificial Life: A Hybrid of the European Conference on Artificial Life (ECAL) and the International Conference on the Synthesis and Simulation of Living Systems (ALIFE) (30), pp. 444–451 (2018)
4. Borchert, R., Honda, H.: Control of development in the bifurcating branch system of Tabebuia rosea: a computer simulation. Bot. Gaz. **145**(2), 184–195 (1984)
5. Boudon, F., Pradal, C., Cokelaer, T., Prusinkiewicz, P., Godin, C.: L-Py: an L-system simulation framework for modeling plant architecture development based on a dynamic language. Front. Plant Sci. **3**, 76 (2012)
6. Chan, W.L., et al.: Wind loading on scaled down fractal tree models of major urban tree species in Singapore. Wind Impacts on Forests and Trees in a Changing Climate – A Special Issue in Collaboration with the IUFRO Working Party 8.03.06 (2020)
7. Paine, C.E.T., et al.: How to fit nonlinear plant growth models and calculate growth rates: an update for ecologists. Methods Ecol. Evol. **3**, 245–256 (2012)
8. Gobeawan, L., et al.: Modeling trees for virtual Singapore: from data acquisition to CityGML models. Int. Arch. Photogram. Remote Sens. Spat. Inf. Sci. **XLII-4/W10**, 55–62 (2018)
9. Gobeawan, L., et al.: Species-level tree biomechanical models. In: 9th IUFRO Wind and Trees Conference (2020)
10. Gobeawan, L., et al.: Convenient tree species modeling for virtual cities. In: Gavrilova, M., Chang, J., Thalmann, N.M., Hitzer, E., Ishikawa, H. (eds.) CGI 2019. LNCS, vol. 11542, pp. 304–315. Springer, Cham (2019). https://doi.org/10.1007/978-3-030-22514-8_25
11. Godin, C., Costes, E., Sinoquet, H.: A method for describing plant architecture which integrates topology and geometry. Ann. Bot. **84**(3), 343–357 (1999). http://www.jstor.org/stable/42766003
12. Godin, C., Sinoquet, H.: Functional-structural plant modelling. New Phytol. **166**(3), 705–708 (2005)
13. Hallé, F., Oldeman, R.A.A., Tomlinson, P.B.: Tropical Trees and Forests: An Architectural Analysis. Springer, New York (1978). https://doi.org/10.1007/978-3-642-81190-6
14. Hu, S., Li, Z., Zhang, Z., He, D., Wimmer, M.: Efficient tree modeling from airborne lidar point clouds. Comput. Graph. **67**(C), 1–13 (2017)
15. Joo, P.H., et al.: Wind load prediction on single tree with integrated approach of L-system fractal model, wind tunnel, and tree aerodynamic simulation. AIP Adv. **10**(6) (2020)
16. Kamal, M., Phinn, S., Johansen, K.: Object-based approach for multi-scale mangrove composition mapping using multi-resolution image datasets. Remote Sens. **7**(4), 4753–4783 (2015)

17. Kang, M., Hua, J., De Reffye, P., Jaeger, M.: Parameter identification of plant growth models with stochastic development, pp. 98–105 (2016)
18. Kass, R.E., Carlin, B.P., Gelman, A., Neal, R.M.: Markov chain monte Carlo in practice: a roundtable discussion. Am. Stat. **52**(2), 93–100 (1998)
19. Lim, C.W., et al.: Generation of tree surface mesh models from point clouds using skin surfaces. In: 15th International Conference on Computer Graphics Theory and Applications, pp. 83–92 (2020)
20. Lin, E.S., Teo, L.S., Yee, A.T.K., Li, Q.H.: Populating large scale virtual city models with 3D trees. In: 55th IFLA (International Federation of Landscape Architects) World Congress (2018)
21. Lin, E.S., Gobeawan, L., Yee, A.T.K., Wong, S.T., Poto, M., Tandon, A.: PlantXML: a data exchange format for tree information. In: 2nd International IAG Workshop on BIM and GIM Integration (2019)
22. Lindenmayer, A.: Mathematical models for cellular interactions in development I. Filaments with one-sided inputs. J. Theor. Biol. **18**(3), 280–299 (1968)
23. Lintermann, B., Deussen, O.: Interactive modeling of plants. IEEE Comput. Graph. Appl. **19**(1), 56–65 (1999)
24. Livny, Y., et al.: Texture-lobes for tree modelling. ACM Trans. Graph. **30**(4), 53:1–53:10 (2011)
25. Longay, S., Runions, A., Boudon, F., Prusinkiewicz, P.: TreeSketch: interactive procedural modeling of trees on a tablet. In: Kara, L., Singh, K. (eds.) EUROGRAPHICS Symposium on Sketch-Based Interfaces and Modeling. Cagliari, Italy (2012)
26. Makowski, M., Hädrich, T., Scheffczyk, J., Michels, D.L., Pirk, S., Pałubicki, W.: Synthetic silviculture: multi-scale modeling of plant ecosystems. ACM Trans. Graph. **38**(4) (2019). https://doi.org/10.1145/3306346.3323039
27. McKay, M.D.: Latin hypercube sampling as a tool in uncertainty analysis of computer models. In: Proceedings of the 24th Conference on Winter Simulation, WSC 1992, pp. 557–564. ACM, New York (1992)
28. Mitchell, M.: An Introduction to Genetic Algorithms. MIT Press, Cambridge (1998)
29. National Research Foundation Singapore: Virtual Singapore. https://www.nrf.gov.sg/programmes/virtual-singapore
30. Niese, T., Pirk, S., Albrecht, M., Benes, B., Deussen, O.: Procedural Urban Forestry (2020)
31. Prusinkiewicz, P., Lindenmayer, A.: The Algorithmic Beauty of Plants. Springer, Heidelberg (1996)
32. Reche-Martinez, A., Martin, I., Drettakis, G.: Volumetric reconstruction and interactive rendering of trees from photographs. ACM Trans. Graph. **23**(3), 720–727 (2004)
33. Sagar, M., Miranda, J., Dhawan, V., Dharmaraj, S.: The growing trend of city-scale digital twins around the world (2020). https://opengovasia.com/the-growing-trend-of-city-scale-digital-twins-around-the-world/
34. Sievänen, R., Godin, C., Dejong, T., Nikinmaa, E.: Functional-structural plant models: a growing paradigm for plant studies. Ann. Bot. **114**, 599–603 (2014)
35. Stava, O., et al.: Inverse procedural modelling of trees. Comput. Graph. Forum **33**(6), 118–131 (2014)
36. Talle, J., Kosinka, J.: Evolving L-systems in a competitive environment. In: Magnenat-Thalmann, N., et al. (eds.) CGI 2020. LNCS, vol. 12221, pp. 326–350. Springer, Cham (2020). https://doi.org/10.1007/978-3-030-61864-3_28

37. Vos, J., Evers, J.B., Buck-Sorlin, G.H., Andrieu, B., Chelle, M., de Visser, P.H.B.: Functional-structural plant modelling: a new versatile tool in crop science. J. Exp. Bot. **61**(8), 2101–2115 (2010)
38. Weber, J., Penn, J.: Creation and rendering of realistic trees. In: Proceedings of the 22nd Annual Conference on Computer Graphics and Interactive Techniques, SIGGRAPH 1995, pp. 119–128. ACM, New York (1995)
39. Xu, H., Gossett, N., Chen, B.: Knowledge and heuristic-based modeling of laser-scanned trees. ACM Trans. Graph. **26**(4) (2007)
40. Xu, L., Mould, D.: Procedural tree modeling with guiding vectors. Comput. Graph. Forum **34**(7), 47–56 (2015)
41. Yi, L., Li, H., Guo, J., Deussen, O., Zhang, X.: Tree growth modelling constrained by growth equations. Comput. Graph. Forum **37**(1), 239–253 (2018)

Self-restructuring of Mesh-Connected Processor Arrays with Spares Assigned on Rotated Orthogonal Side

Itsuo Takanami[1]([✉]), Masaru Fukushi[2], and Takahiro Watanabe[3]

[1] Department of Technology, Yamaguchi University, Tokiwadai 2-16-1,
Ube 755-8611, Japan
iftakanami@comet.ocn.ne.jp
[2] Graduate School of Sciences and Technology for Innovation, Yamaguchi University,
Tokiwadai 2-16-1, Ube 755-8611, Japan
mfukushi@yamaguchi-u.ac.jp
[3] Graduate School of Information, Production and Systems, Waseda University,
2-7 Hibikino, Wakamatsu, Kitakyushu-shi 808-0135, Japan
watt@waseda.jp

Abstract. We propose a new restructuring method for mesh-connected processor arrays (PAs) using single track shift with spare processing elements (PEs) on the orthogonal two or four sides. First, a method called "ROT method" for mesh-connected PAs with spare PEs on the orthogonal sides is described. Next, the method is applied to mesh-conected PAs with spare PEs on the four sides. The effect of the method is evaluated by computer simulation and it is shown that ROT method makes array reliabilites fairly higher than those of the existing methods for mesh-connected PAs with the same number of spare PEs and the same network structure. The hardware realization of ROT method is also discussed and it is seen that the method can be realized as a simple built-in circuit. The circuit can be embedded in a target PA to reconstruct the PA with faulty PEs very quickly. This implies that the proposed method is especially effective in enhancing the run-time reliability of a PA in mission critical systems where first self-reconfiguration is required without an external host computer or manual maintenance operations.

Keywords: Fault-tolerance · Mesh array · Single track shift · Self-restructuring · Built-in circuit

1 Introduction

In recent years, there has been a rapidly growing interest in processing many kinds of vast amount of information in real-time and near-real-time. The demand for strengthening computation power will never stop, and it is increasing day by day. For these needs, how to realize high-speed and massively parallel computers has been the subject of extensive research. A mesh-connected processor array

© Springer-Verlag GmbH Germany, part of Springer Nature 2021
M. L. Gavrilova and C. J. K. Tan (Eds.): Trans. on Comput. Sci. XXXVIII, LNCS 12620, pp. 36–53, 2021.
https://doi.org/10.1007/978-3-662-63170-6_3

(PA) is a kind of form of massively parallel computers. Mesh-connected PAs consisting of hundreds of processing elements (PEs) offer a regular and modular structure, a small wiring length between PEs, and a high scalability, thus very suitable for most signal and image processing algorithms.

One of the most important and fundamental issues that must be addressed for such PAs is defect/fault tolerance. If a single PE fails to perform its assigned task correctly due to some defects/faults, the entire computation will result in failure. On the other hand, as VLSI technology has developed, the realization of parallel computing systems using multi-chip module (MCM) e.g., [1], wafer scale integration (WSI) e.g., [2] or network-on-chip (NoC) e.g., [3] has been considered so as to enhance the computation and communication performance, decrease energy consumption and sizes, and so on. In such a realization, entire or significant parts of PEs and interconnections among them are implemented on a single chip or wafer. Therefore, the yield and/or reliability of the system may become drastically low if no strategy is employed for coping with defects and faults.

To restore the correct computation capabilities of PAs, it must be reconfigured appropriately so that defective PEs are eliminated from the computation paths and the remaining PEs maintain correct logical connectivities between themselves, i.e., keep a logical mesh structure. Various strategies to reconfigure a faulty physical system into a fault-free target logical system are described in the literature, e.g., [4–9]. These strategies are categorized in the redundancy approach. In contrast, there are those in the degradation approach in which no spare PEs are designated in advance and a reconfiguration is done by avoiding faulty PEs with keeping the array structure, e.g., [10–13]. Some of these techniques employ very powerful reconfiguring systems that can repair a faulty PA with almost certainty, even in the presence of clusters of multiple faults. However, the key limitation of these techniques is that they are executed in software programs to run on an external host computer and they cannot be designed and implemented efficiently within a PA chip as dedicated circuits. If a faulty PA can be reconfigured by a built-in circuit, the system down time of the PA is significantly reduced. Furthermore, the PA will become more reliable when it is used in such an environment where the fault information cannot be monitored externally through the boundary pins of the chip and manual maintenance operations are difficult.

As far as we know, the first attempts to develop a restructuring scheme using hardware switching were made by Negrini et al. [14] and Sami and Stefanelli [15]. The design approach of their repair methods begins with a heuristic algorithm that involves only some local reconnection operations and is not for self-restructuring. Further, the interconnection structure used multiplexers inserted between logically adjacent PEs, which may cause large communication delays between PEs.

Takanami et al. developed the automatic self-restructuring schemes (SRSs) [16–19] for mesh-connected PAs using single track switches proposed by Kung et al. [4]. The arrays they dealt with have two rows and two columns of spares, i.e., four linear arrays of spares around a PA to cope with faults due to low reliable PEs. However, the hardware complexity and restructuring time is not so small. Besides the above approaches, Lin et al. [20] proposed the fault-tolerant path router with built-in self-test/self-diagnosis and fault-isolation circuits for 2D-mesh based chip multiprocessor systems. Collet et al. [21] also proposed the chip self-organization and fault tolerance at the architectural level to improve dependable continuous operation of multicore arrays in massively defective nanotechnologies, where the architectural self-organization results from the conjunction of self-diagnosis and self-disconnection mechanism, plus self-discovery of routers to maintain the communication in the array.

On the other hand, if PEs are fairly reliable, it is expected that the smaller number of spares will be sufficient for retaining the reliability of an array so high and additional control schemes as well as control circuits will become simple. From this expectation, Takanami et al. proposed SRSs for PAs with spares on one row and one column [22, 23]. In these schemes, $2N$ spares are used for arrays with size of $N \times N$.

This paper is a revised and extended version of [24], in which the basic idea of rotating the assignment of spares on the sides of arrays was introduced and it was shown that the array reliabilities increased, comparing with those of the non-rotating method. In this paper, a hardware realization (i.e., built-in circuit) for the introduced restructuring method (ROT method) with rotating the assignment of spares is described in detail. Next, it is discussed to apply ROT method to an array with spares on it's four sides by dividing the array into four subarrays which is denoted as "4ss-array", considering whether the connections among the subarrays after rotating the assignment of spares can reconstruct the logical mesh array structure. As the result, it is shown that the application of ROT method to 4ss-arrays is effective, if the hardware realization is considered, To further increase the array reliability using ROT method, a method which changes dividing positions variably is introduced. Then, it is shown that this method increases the reliabilities of arrays to be restructured extremely so much.

The paper is organized as follows. In Sect. 2, a new method which is called ROT method for mesh-connected PAs with spare PEs on the orthogonal two sides is described. In Sect. 3, the array reliabilities of ROT method are shown by computer simulations, comparing with those of the existing methods. In Sect. 4, a logical circuit to realize ROT method in hardware is described. In Sect. 5, methods to apply ROT method to the arrays with spares on the four sides are presented, where they are divided into four subarrays and ROT method is applied to each of them. The array reliabilities and hardware realization are also described. Section 6 is the conclusion.

2   Orthogonal Spare Side Rotation

2.1   Orthogonal Spare Side Scheme (OTSS)

Figure 1 shows the arrangement of spare PEs in OTSS where linear arrays of spares are on the 0-th row and 0-th column of an array with a size of $(M + 1) \times$

$(N + 1)$. Here, the example case of $M = 3$ and $N = 4$ is shown in Fig. 1. Then, restructuring is done so that an array with a logical size of $M \times N$ consisting of healthy PEs is obtained. For this arrangement, several network architectures can be considered according to how an array is restructured using spares. Here, we adopt the single track shift method [4] in which the restructuring is done by shifting PEs. It has an advantage that the physical distances among logically adjacent PEs after restructuring are bounded by a constant.

Figure 2 shows the network structure for OTSS using single track shift. This is a mesh-connected PA proposed by Kung et al. [4] except that it has only one spare row and one spare column. Tracks consisting of links and switches run along toward the spares located on the upper and left sides. A single track runs between adjacent rows and columns, and there is a switch at the crosspoint of a track and a link connecting PEs.

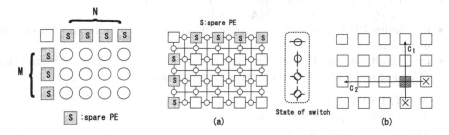

Fig. 1. An arrangement
of spares in OTSS.

Fig. 2. (a) OTSS using single track shift.
(b) c-path c_1 and c_2 in intersection relation.

The replacement of a faulty PE is done as follows. A faulty PE marked with $\times$ is bypassed in the horizontal or vertical direction and replaced by its adjacent healthy PE, which in turn is replaced by the next adjacent healthy PE, and so on. This replacement is repeated until a healthy spare PE is used in the end. This process defines a compensation path (shortly written as c-path), which is a set of straight and continuous PEs from a faulty PE to a healthy spare PE.

Notation:

- c-paths c_1 and c_2 are called to be in intersection relation if they include the same PE as shown in Fig. 2(b).
- This replacement method is called "replacement scheme with spares on orthogonal side using single track shift" and shortly written as RS-OTSTS.

For each faulty nonspare PE, there may be two straight c-paths which go in the left and upper directions. Then, we have the following property.

Property 1. (Restructurability condition (RC) for RS-OTSTS) An array with faulty PEs is restructurable if and only if there is a set of c-paths S such that

1. S contains a c-path for each nonspare faulty PE, and
2. No c-paths in S are in intersection relation.   □

We present a restructuring algorithm (RA-OTSTS) by which we can judge whether the condition of Property 1 is satisfied for an array with faulty PEs, where the array size is $(M + 1) \times (N + 1)$.

Notation:

- $PE(i, j)$ $(0 \le i \le M, 0 \le j \le N)$ denotes the PE at physical location (i, j).
- Spares are assigned on the upper and left sides; that is, at $(0, j)$'s $(1 \le j \le N)$ and at $(i, 0)$'s $(1 \le i \le M)$.
- p_{ij} is the fault state of $PE(i, j)$ where $p_{ij} = 1$ and $= 0$ mean that $PE(i, j)$ is faulty and healthy, respectively.
- A matrix $P = (p_{ij})$ is called a fault pattern.
- $D = (d_{ij})$ is a matrix with the same size as that of P.
- flg and sfg are flags used in RA-OTSTS.
- If a faulty PE is replaced by a spare PE, it is said to be repaired, otherwise unrepaired.
- For a fault pattern P, if all the faults in P can be repaired at the same time, P is said to be repairable, otherwise irreparable.

RA-OTSTS
Input: a fault pattern $P = (p_{ij})$.
Output: the value of sfg.
Do the following Steps
    Step 1. Set the values of all the elements in D to 0s and the value of sfg to 1.
    Step 2. Decreasing the value of the variable j from N to 1, do the following.
        Step 2-1. Set the value of flg to p_{0j}.
        Step 2-2. Increasing the value of the variable i from 1 to M,
                repeat Steps 2-2-1 to 2-2-3.
        Step 2-2-1, If $d_{ij}=2$, then set the value of flg to 1.
        Step 2-2-2, If $p_{ij}=1$ and the value of $flg=1$,
                then do Steps 2-2-2-1 to 2-2-2-2.
            Step 2-2-2-1. Set the value of d_{ij} to 2.
            Step 2-2-2-2. Decreasing the value of the variable k from $j - 1$ to 0,
                    do Step 2-2-2-2-1.
                Step 2-2-2-2-1. If $p_{ik}=1$, then set the value of sfg to 0,
                        else the value of d_{ik} to 2.
        Step 2-2-3, If $p_{ij}=1$ and the value of $flg=0$,
                then do Steps 2-2-3-1 to 2-2-3-2
            Step 2-2-3-1. Set the value of flg to 1.
            Step 2-2-3-2. Decreasing the value of the variable k from i to 1,
                    set the value of d_{kj} to 1.
We outline RA-OTSTS.
From the algorithm, we can see the following.

- If $p_{ij} = 1$ and $flg = 1$, a sequence of $d_{ik} = 2$'s ($k \le j$) is generated in the left direction from PE(i, j) toward the spare on the left side in the processes from Steps 2-2-2-1 to 2-2-2-2 in the algorithm. This sequence which reaches the spare is called 2-sequence from PE(i, j), which is denoted as 2-seq(i, j).
- If $p_{ij} = 1$ and $flg = 0$, a sequence of $d_{kj} = 1$'s is generated in the upper direction from PE(i, j) toward the spare on the upper side in Step 2-2-3-2, This sequence is called 1-sequence from PE(i, j), which is denoted as 1-seq(i, j).
- 1- and 2-sequences are candidate c-paths.

We have the following property for restructurability.

Property 2. An array of physical size $(M + 1) \times (N + 1)$ with a fault pattern $P = (p_{ij})$ is repairable into an array of logical size $M \times N$ (in the meaning of Property 1) if and only if RA-OTSTS ends with $sfg = 1$. If it ends by returning $sfg = 1$, the set of 1- and 2-sequences is a set of c-paths which satisfies the RC for RS-OTSTS of Property 1.
Proof: See [22]. □

2.2 Spare Side Rotation Method

The orthogonal spare side rotation method (shortly written as ROT method) to be proposed uses RA-OTSTS as a base algorithm. While in RA-OTSTS spares are fixed on the left and upper sides of an array, in ROT method they are flexibly assigned on one of the four orthogonal sides which are the left and upper, the upper and right, the right and lower, and the lower and left sides. These assignments are denoted as Rot-0, Rot-90, Rot-180 and Rot-270, respectively as shown in Fig. 3.

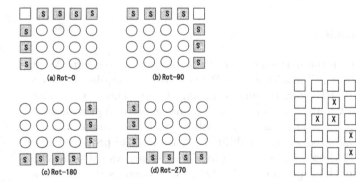

Fig. 3. Four assignments of spares on orthogonal sides.

Fig. 4. A fault pattern which is repairable in Rot-180 but not the other Rot-*'s.

ROT method is done as follows. RA-OTSTS is modified according to each of the four assignments mentioned above where the modification is defined below as in Eq. (1). Then, each of the modified RA-OTSTS's is applied at a time to an array with faults and if at least one of them ends with $sfg = 1$, the array with faulty PEs is judged to be repairable and the array is called to be R-repairable.

Instead of the above, one to fix the spares on one of the four orthogonal sides is called nonROT method.

Figure 4 shows a fault pattern which is repairable for Rot-180 but not the other Rot-∗'s. That is, though the pattern is not repairable for the statically fixed spare assignments as in Rot-0, Rot-90 and Rot-270, it is R-repairable. This suggests that ROT method may increase array reliabilities.

The restructuring algorithm for ROT method is given by the coordinate transformation below to the variables i and j in RA-OTSTS (which is one of nonROT methods) and replacing p_{ij} and d_{ij} with p_{xy} and d_{xy}, respectively.

The coordinate transformation is given by

$$\begin{pmatrix} x \\ y \end{pmatrix} = \begin{pmatrix} b_1 \\ b_2 \end{pmatrix} + \begin{pmatrix} a_{11} & a_{12} \\ a_{21} & a_{22} \end{pmatrix} \begin{pmatrix} i \\ j \end{pmatrix} \tag{1}$$

The parameters $b_1, b_2, a_{11}, ..., a_{22}$ are determined for each of Rot-0, -90, -180 and -270, as in Table 1, for $N = M$.

Table 1. The parameters $b_1, b_2, a_{11}, ..., a_{22}$ for the respective rotations.

	b_1	b_2	a_{11}	a_{12}	a_{21}	a_{22}
Rot-0	0	0	1	0	0	1
Rot-90	0	N	0	1	-1	0
Rot-180	M	N	-1	0	0	-1
Rot-270	M	0	0	-1	1	0

3   Array Reliabilities

In this section, to estimate the performance of ROT method, we execute Monte Carlo simulations, using a PC machine with Borland C++Compiler 5.5. Here, it is assumed that all the PEs may become uniformly faulty. Then, 10^6 random fault patterns each with k faulty PEs for $1 \leq k \leq N_s$ are generated provided that each PE in an array has the same reliability p, where N_s is the number of spare PEs. Then, we have evaluated array reliabilities (ARs) for the proposed ROT method, comparing with those of the existing methods for arrays with the same network structure. Here, the array reliability $AR(p)$ is defined as the sum of all the probabilities each of which is computed as the product of probability that a fault pattern is repaired and one that the pattern occurs under the condition that each PE may be healthy with equal probability p, as below. Then, $AR(p)$ is given as

$$AR(p) = \sum_{k=0}^{N_s} {}_{Na}C_k \cdot SV(k) \cdot p^{N_a - k} \cdot (1 - p)^k,$$
$$SV(k) = \frac{N_{rep}(k)}{N_{pat}(k)},$$

where $N_{rep}(k)$, $N_{pat}(k)$, and N_a are the number of fault patterns which have k faulty PEs and are judged to be repairable, the number of examined fault patterns which have k faulty PEs, and the number of all PEs, respectively. Note that $p^{N_a-k} \cdot (1-p)^k$ is the probability that a fault pattern with k faulty PEs occurs, $_{N_a}C_k = \frac{N_a!}{(N_a-k)!k!}$ the number of fault patterns each with k faulty PEs, $SV(k)$ so called the survival rate which is the ratio of the number of repairable fault patterns each with k faulty PEs and the number of fault patterns each with k faulty PEs examined, which is the probability estimated by simulation that fault patterns each with k faulty PEs are repaired.

Figure 5 shows the survival rates of arrays with size of 16×16.

Figures 6 and 7 show the ARs with sizes of 8×8 and 16×16, respectively.

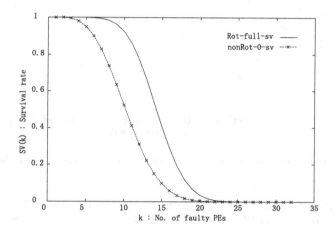

Fig. 5. Survival rate of 16×16 mesh arrays for Rot- and nonRot- methods.

It is seen that the ARs for ROT method (denoted as Rot-full) fairly increase, comparing with those for nonROT method (nonRot-0) and OTSTS-cent+1 and -+2, where OTSTS-cent+1 and -+2 are shown for comparison. Here, OTSTS-cent+1 and -+2 are the ARs for the variants of OTSTS in terms of spare placement as shown in Fig. 8. OTSTS-cent+1 is the result when RA-OTSTS is applied once in order of 1st, 2nd, 3rd and 4th quadrants. OTSTS-cent+2 is the result when RA-OTSTS is applied four times where it is applied the first time in order of 1st, 2nd, 3rd and 4th quadrants, the second time in order of 2nd, 3rd, 4th and 1st quadrants, the third time in order of 3rd, 4th, 1st and 2nd quadrants, and the fourth time in order of 4th, 1st, 2nd and 3rd quadrants.

4 Hardware Realization

4.1 Hardware Realization of RA-OTSTS

We have seen that ROT method has the highest efficiency. So, first we will show a hardware realization of RA-OTSTS, i.e., a built-in circuit and then, that of

ROT method which is realized with small modification by additing the rotation operation to that of RA-OTSTS. It can autonomously decide whether any fault pattern is repairable, i.e., the circuit outputs a signal indicating that an array is repairable if the pattern satisfies RC for RA-OTSTS, and otherwise, a signal indicating that it is irreparable.

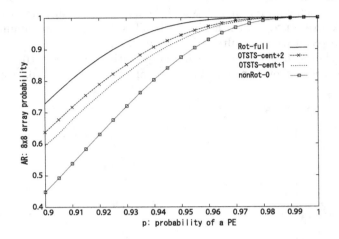

Fig. 6. Reliabilities of 8×8 mesh arrays for Rot- and nonRot- methods.

Figure 9 shows a hardware realization of RA-OTSTS. Each PE has a logical circuit CC shown in Fig. 9(a), which is connected as shown in Fig. 9(b), along with the signals to and from the CCs. The CC has four input signals p, D_{in}, L_{in} and U_{in} and four output signals D_{out}, L_{out}, U_{out} and S_{fg}. The signal p's correspond to p_{ij}'s in RA-OTSTS, D_{in}'s to flg's, L_{out}'s to 2-sequences, U_{out}'s to 1-sequences, respectively, and the logical AND of S_{fg}'s corresponds to sfg. Here, the signals are defined by the equations below.

$$D_{out} = p + D_{in} + L_{in} \qquad (2)$$

$$L_{out} = L_{in} + p \cdot D_{in} \qquad (3)$$

$$U_{out} = U_{in} + p \cdot \overline{D}_{in} \qquad (4)$$

$$\overline{S}_{fg} = p \cdot L_{in} \qquad (5)$$

From the equations above, we can see that the logical circuit of a CC is so simple that the numbers of logical gates and pins are at most ten, respectively.

It has already been shown [22] that the behavior of the CC network shown in Fig. 9 is compatible with that of RA-OTSTS (see Properties 5, 6 and 7). Property 5 is used to check whether an array with faulty PEs is repairable or not. Property 6 shows the correspondence between 2-sequences and sequences of

L_{out}'s. Property 7 shows the correspondence between 1-sequences and sequences of U_{out}'s.

For convenience of explanation, the terminal names and the signals from/in the terminals are often identically used unless confused, and the signals from/in the CC of PE(x, y) are denoted with index '(x, y)'.

Note that $L_{in}(x, N) = 0$ for any x $(1 \le x \le M)$ and $U_{in}(M, y) = 0$ for any y $(1 \le y \le N)$.

Though the CC network shown in Fig. 9(b) behaves in asynchronous mode, it has already been shown [22] that the signals in the network become stable for any fixed input values, which is stated as follows.

Property 3. Let t_d denote a signal propagation time in a CC. For any column of CCs in an array of logical size $N \times N$, the signals in the column become stable within $(2N - 1) \cdot t_d$ after the external input values to the column are fixed. $\square$

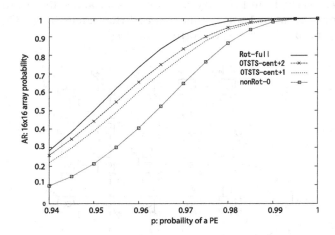

Fig. 7. Reliabilities of 16×16 mesh arrays for Rot- and nonRot- methods.

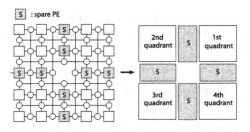

Fig. 8. A variant in terms of spare replacement where the row and column of spares are placed in the center of an array.

Since the external signals on a column come from its right neighbors, we have the following properties [22].

Property 4. For the CC network for an array of logical size $N \times N$, the signals become stable within $(2N-1)N \cdot t_d$ for any fault pattern. □

In the following, from Property 4, we assume that the signals have become stable. Now, we describe the following properties which are closely related to the hardware realization of RA-OTSTS.

Property 5. An array with faulty PEs is repairable, that is, RC-OTSTS is satisfied if and only if all the output signals S_{fg}'s from CCs are 1s. □

Now, we explain how to switch the connections among PEs if the logical AND of all the output signals S_{fg}'s from CCs is 1. Hence, in the following, we assume that a repairable fault pattern is given, that is, the logical AND of all the output signals S_{fg}'s from CCs is 1. In order to switch the connections among PEs correctly, from Property 2, it is necessary and sufficient to know the correspondences between 1- or 2-sequences and the signals in the CC network. The following properties give such correspondences.

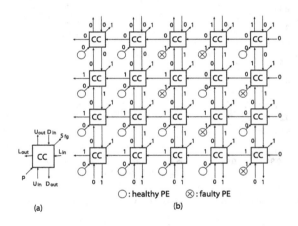

(a)   (b)

○ : healthy PE   ⊗ : faulty PE

Fig. 9. CC network to realize RA-OTSTS where (a) Input and output signals of control circuit CC. (b) Connection of the CC and an illustration of inputs and outputs to/from CCs for a fault pattern.

Property 6. $d_{xy} = 2$ if and only if $L_{out}(x, y) = 1$.
□

Property 7. $d_{xy} = 1$ if and only if $U_{out}(x, y) = 1$. □

From the results above, it is seen that there are one-to-one correspondences between 1-sequences and sequences of U_{out}'s, and between 2-sequences and sequences of L_{out}'s. Hence, from Property 2, we can easily give a truth table defining the states of switches, using U_{out}'s and L_{out}'s. However, since it is a simple task, we will omit it.

4.2   Hardware Realization of ROT Method

To realize ROT method in hardware, it is sufficient to rotate the CCs in Fig. 9 for the respective rotation degrees, i.e., 90°, 180° and 270°, and properly give the signal 0s to the CCs on the sides of the CC network. Figures 10 and 11 show how to rotate them by adding sw's and lines around the CC and changing the switch states of the sw's as shown, according to the respective degrees. This CC modified is denoted as ex-CC and the CC network composed of ex-CCs is denoted as ex-CC network. Figure 12 shows how to give the signal 0s to the sides of ex-CC network for each rotation.

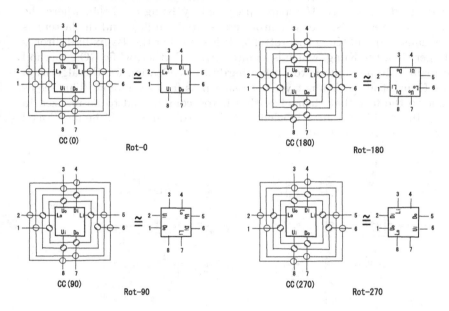

Fig. 10. States of sw's to rotate CC in CC network for the degrees 0° and 90°. CCs rotated are denoted as ex-CC(0) and ex-CC(90), respectively.

Fig. 11. States of sw's to rotate CC in CC network for the degrees 180° and 270°. CCs rotated are denoted as ex-CC(180) and ex-CC(270), respectively.

5   Application to Array with Spares on Four Sides

We have investigated the case that the spares are on two sides of an $(N + 1) \times (N + 1)$ array in which the number of the spares is $2N + 1$. Of course, it is necessary to increase the number of spares to cope with the case that each PE is not so reliable. A method to increase the number of spares using RA-OTSTS is to divide a given array into subarrays each of which RA-OTSTS (nonROT method) or ROT method is applied to.

5.1   Fixed Divide Case

For example, an $(N+2) \times (N+2)$ array with spares on the four sides shown in Fig. 13 is divided into four subarrays as shown in Fig. 14, each with the size of $(\frac{N}{2}+1) \times (\frac{N}{2}+1)$. RA-OTSTS or ROT method is applied to each subarray, where the $(N+2) \times (N+2)$ array (denoted shortly as 4ss-array) has $(4N+4)$ spares. Then, a 4ss-array with faulty PEs is judged to be repairable if and only if all the subarrays are repairable by ROT method. The AR of the 4ss-array by ROT method is calculated by the 4-th power of that of the subarray. Figure 15 shows the case of $N = 16$, comparing the ARs of the arrays in OTSTS or ROT method with that of an array with $4N$ spares proposed by Kung et al. [4], where the formers are denoted as 4ss-fixdiv-nonrot and 4ss-fixdiv-fullrot and the latters as Kung (exhaustive) and Kung (neural) [25]. It is seen that the 4ss-fixdiv-fullrot is fairly larger than the Kung, which is so interesting. Further, while ROT method can be realized in a simple hardware, a hardware realization for the algorithm given by Kung et al. [4] has not yet been shown as far as we know. This seems to be due to the fact that their algorithm is too complicated to be realized in hardware. In Fig. 15, the curve labeled with 4ss-vardiv-fullrot will be mentioned in the following variable divide case.

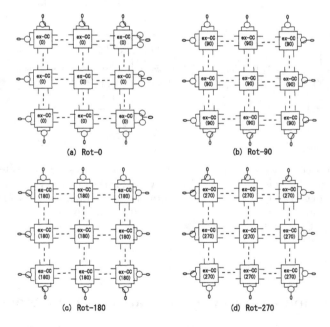

Fig. 12. States of sw's on the sides of ex-CC network for Rot-0, -90, -180 and -270, respectively.

To realize ROT method in hardware, it must be shown that subarrays neighboring with each other after rotations can be connected, keeping the neighboring relation among logical addresses, by switching sw's, where spares assigned are allowed to be passed. Figure 16 shows an illustration of connections between the cases that the subarrays of Rot-0 and Rot-180 are adjacent to each other. It is easily seen that the connections are surely done, keeping the neighboring relation among logical addresses. Note that unused spare PEs are allowed to be passed vertically and horizontally, using internal switches, as commonly assumed in the existing methods. It could also be shown that the connections between other Rot-* cases can surely be done, but it is omitted to illustrate them because it is a simple but tedious task.

5.2   Variable Divide Case

We will further discuss the case that a 4ss-array is variably divided into four subarrays, i.e., row and column positions to be divided are variable. To simulate the case, RA-OTSTS is modified as follows.

Let the physical addresses at the top and leftmost and the bottom and rightmost be (x_0, y_0) and (x_1, y_1) $(x_0 \leq x_1, y_0 \leq y_1)$, respectively. When RA-OTSTS

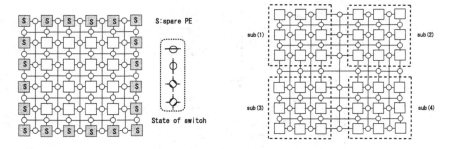

Fig. 13. An array with spares on four sides.

Fig. 14. 4ss-array divided into four subarrays.

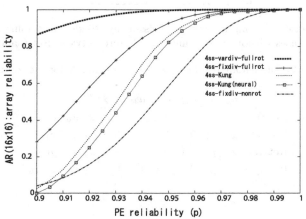

Fig. 15. Reliabilities of 4ss-arrays with size of 16×16.

is applied to this subarray, the coordinate transformations for RA-OTSTS are done in accordance with the parameters in Table 2 as in Eq. (1), and p_{ij} and d_{ij} are replaced by p_{xy} and d_{xy}, respectively. Further M and N in RA-OTSTS are substituted by the parameters in Table 2.

Figure 17 shows the ex-CC network for a 4ss-array which is modified so that the array to be restructured can be divided at arbitrary row and column positions (refer to Fig. 12 about how to switch sw's to input 0s on the sides of ex-CCs). Then, the array is judged to be repairable if and only if all the subarrays are R-repairable at some row and column positions divided. This method is denoted as varROT method. Figure 17 illustrates the case that the array is divided between the third and fourth rows and between the fourth and fifth columns.

The label "4ss-vardiv-fullrot" in Fig. 15 shows the AR of the array with size of 16×16 where "4ss-fixdiv fullrot" corresponds to the fixed divide case. From the figure, though the additional sw's and a switch control scheme are needed, it is seen that the ARs by ROT methods increase, and extremely so that by varROT method.

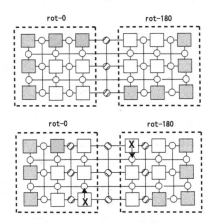

Fig. 16. Connections among subarrays of 4ss-array divided into four subarrays. (a) and (b) are the cases without failure and with failure between Rot-0 and Rot-180.

Table 2. The parameters $b_1, b_2, a_{11}, ..., a_{22}, M, N$ for the rotation.

	b_1	b_2	a_{11}	a_{12}	a_{21}	a_{22}	M	N
Rot-0	x_0	y_0	1	0	0	1	x_1-x_0	y_1-y_0
Rot-90	x_0	y_1	0	1	−1	0	y_1-y_0	x_1-x_0
Rot-180	x_1	y_1	−1	0	0	−1	x_1-x_0	y_1-y_0
Rot-270	x_1	y_0	0	−1	1	0	y_1-y_0	x_1-x_0

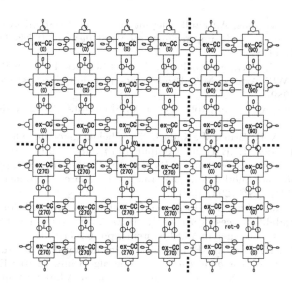

Fig. 17. 4ss-array variably divided into four subarrays.

6 Conclusion

We have presented ROT method for restructuring a mesh-connected PA with spare PEs. Then, it has been shown that the array reliabilities by ROT method fairly increase, comparing with those by nonROT and the existing methods with the same number of spares and the same network structure. Further, it has been shown that ROT method can be realized as a simple built-in circuit. The failure of the built-in circuit can also be considered to be negligible since the hardware complexity of the CC or ex-CC network is usually much less than that of an array. Hence, the proposed method is especially effective in enhancing the run-time reliability of a PA in mission critical systems where first self-reconfiguration is required without an external host computer or manual maintenance operations.

Acknowledgment. The research is in part supported by Yamaguchi University Fund.

References

1. Schaper, L.W.: Design of multichip modules. Proc. IEEE **80**(12), 1955–1964 (1992)
2. Okamoto, K.: Importance of wafer bonding for the future hype-miniaturized CMOS devices. ECS Trans. **16**(8), 15–29 (2008)
3. Okamoto, K.: Route packets, not wires: on-chip interconnection networks. In: Proceedings of the 38th Design Automation Conference, pp. 684–689 (2001)
4. Kung, S.Y., Jean, S.N., Chang, C.W.: Fault-tolerant array processors using single-track switches. IEEE Trans. Comput. **38**(4), 501–514 (1989)

5. Mangir, T.E., Avizienis, A.: Fault-tolerant design for VLSI: effect of interconnection requirements on yield improvement of VLSI designs. IEEE Trans. Comput. **31**(7), 609–615 (1982)
6. Negrini, R., Sami, M.G., Stefanelli, R.: Fault-Tolerance Through Reconfiguration of VLSI and WSI Arrays. MIT Press Series in Computer Systems. MIT Press, Cambridge (1989)
7. Koren, I., Singh, A.D.: Fault tolerance in VLSI circuits. IEEE Comput. **23**, 73–83 (1990)
8. Roychowdhury, V.P., Bruck, J., Kailath, T.: Efficient algorithms for reconstruction in VLSI/WSI array. IEEE Trans. Comput. **39**(4), 480–489 (1989)
9. Varvarigou, T.A., Roychowdhury, V.P., Kailath, T.: Reconfiguring processor arrays using multiple-track models: the 3 - tracks - 1 - spare - approach. IEEE Trans. Comput. **42**(11), 1281–1293 (1993)
10. Fukushi, M., Fukushima, Y., Horiguchi, S.: A genetic approach for the reconfiguration of degradable processor arrays. In: IEEE 20th International Symposium on Defect and Fault Tolerance in VLSI Systems, pp. 63–71 (2005)
11. Fukushima, Y., Fukushi, M., Horiguchi, S.: An improved reconfiguration method for degradable processor arrays using genetic algorithm. In: IEEE 21st International Symposium on Defect and Fault Tolerance in VLSI Systems, pp. 353–361 (2006)
12. Wu, J., Zhu, L., He, P., Jiang, G.: Reconfigurations for processor arrays with faulty switches and links. In: 15th IEEE/ACM International Symposium on Cluster, Cloud and Grid Computing, pp. 141–148 (2015)
13. Qian, J., Zhou, Z., Gu, T., Zhao, L., Chang, L.: Optimal reconfiguration of high-performance VLSI subarrays with network flow. IEEE Trans. Parallel Distrib. Syst. **27**(12), 3575–3587 (2016)
14. Negrini, R., Sami, M., Stefanelli, R.: Fault tolerance techniues for array structures used in supercomputing. IEEE Comput. **19**(2), 78–87 (1986)
15. Sami, M., Stefanelli, R.: Reconfigurable architectures for VLSI processing arrays. Proc. IEEE **74**, 712–722 (1986)
16. Takanami, I., Kurata, K., Watanabe, T.: A neural algorithm for reconstructing mesh-connected processor arrays using single-track switches. In: International Conference on WSI, pp. 101–110 (1995)
17. Horita, T., Takanami, I.: An efficient method for reconfiguring the $1\frac{1}{2}$ track-switch mesh array. IEICE Trans. Inf. Syst. **E82–D**(12), 1545–1553 (1999)
18. Horita, T., Takanami, I.: Fault tolerant processor arrays based on the $1\frac{1}{2}$-track switches with flexible spare distributions. IEEE Trans. Comput. **49**(6), 542–552 (2000)
19. Horita, T., Takanami, I.: An FPGA implementation of a self-reconfigurable system for the $1\frac{1}{2}$ track-switch 2-D mesh array with PE faults. IEICE Trans. Inf. Syst. **E83–D**(8), 1701–1705 (2000)
20. Lin, S.Y., Shen, W.C., Hsu, C.C., Wu, A.Y.: Fault-tolerant router with built-in self-test/self-diagnosis and fault-isolation circuits for 2D-mesh based chip multiprocessor systems. Int. J. Electr. Eng. **16**(3), 213–222 (2009)
21. Collet, J.H., Zajac, P., Psarakis, M., Gizopoulos, D.: Chip self-organization and fault-tolerance in massively defective multicore arrays. IEEE Trans. Dependable Secure Comput. **8**(2), 207–217 (2011)
22. Takanami, I.: Self-reconfiguring of $1\frac{1}{2}$-track-switch mesh arrays with spares on one row and one column by simple built-in circuit. IEICE Trans. Inf. Syst. **E87–D**(10), 2318–2328 (2004)

23. Takanami, I., Horita, T., Akiba, M., Terauchi, M., Kanno, T.: A built-in self-repair circuit for restructuring mesh-connected processor arrays by direct spare replacement. In: Gavrilova, M.L., Tan, C.J.K. (eds.) Transactions on Computational Science XXVII. LNCS, vol. 9570, pp. 97–119. Springer, Heidelberg (2016). https://doi.org/10.1007/978-3-662-50412-3_7

24. Takanami, I., Fukushi, M.: Restructuring mesh-connected processor arrays with spares on four sides by orthogonal side rotation. In: Proceedings of 23rd Pacific Rim International Symposium on Dependable Computing, pp. 181–182 (2018)

25. Horita, T., Takanami, I.: A built-in self-reconstruction approach for partitioned mesh-arrays using neural algorithm. IEICE Trans. Inf. Syst. **E79–D**(8), 1160–1167 (1996)

Algorithms for Generating Strongly Chordal Graphs

Asish Mukhopadhyay$^{(\boxtimes)}$ and Md. Zamilur Rahman

School of Computer Science, University of Windsor, Windsor, Canada
{asishm,rahma11u}@uwindsor.ca

Abstract. Graph generation serves many useful purposes: cataloguing, testing conjectures, to which we would like to add that of producing test instances for graph algorithms. Strongly chordal graphs are a subclass of chordal graphs for which polynomial-time algorithms could be designed for problems which are NP-complete for the parent class of chordal graphs. In this paper, we propose three different algorithms for generating strongly chordal graphs, each based on a different characterization of strongly chordal graphs. Each one of them is interesting in its own right, but the third one has turned out to be the most versatile in the sense that it can generate strongly chordal graphs with multiple components and does not require a chordal graph as input.

Keywords: Graph algorithms · Graph generation · Chordal graphs · Strongly chordal graphs

1 Introduction

Strongly chordal graphs are a subclass of the well-studied class of chordal graphs. The interest in this class stems from the fact that many hard problems are solvable in polynomial time for this class of graphs. Evidence of this interest is apparent from the published research, dealing with various aspects of strongly chordal graphs, [1,4,9,12]. In this paper we explore a number of different methods for generating strongly chordal graphs. This would be of interest if we were to test, for example, an implementation of a polynomial-time algorithm for k-tuple dominating sets for strongly chordal graphs [7]. To the best of our knowledge there does not seem to exist any such algorithm in the literature. However, a number of different characterizations of strongly chordal graphs are known. Farber [3], for example, established a number of different characterizations that include one based on totally balanced matrices, another that is based on a class of forbidden induced subgraphs called trampolines, and a third based on the notion of a strong chord. These also include an intersection graph characterization [2] that is analogous to a similar characterization for chordal graphs [5].

We propose generation algorithms based on three of the characterizations listed above. In the first method, our algorithm first generates chordal graphs, using an available algorithm (see [8,11]) and then adds enough edges to make

© Springer-Verlag GmbH Germany, part of Springer Nature 2021
M. L. Gavrilova and C. J. K. Tan (Eds.): Trans. on Comput. Sci. XXXVIII, LNCS 12620, pp. 54–75, 2021.
https://doi.org/10.1007/978-3-662-63170-6_4

it strongly chordal, unless it is already so. The edge additions rely on the characterization that a certain neighborhood matrix of a strongly chordal graph is a totally balanced matrix (this is when the neighborhood matrix does not have $\left[\begin{smallmatrix} 1 & 1 \\ 1 & 0 \end{smallmatrix}\right]$ as a submatrix). The second generation method is based on the forbidden subgraph characterization of strongly chordal graphs. Şeker et al. [14] exploited the intersection graph characterization of chordal graphs to obtain an algorithm for generating them. Here, we propose an algorithm to show that strongly chordal graphs can also be generated using their intersection graph characterization. This is our third method for generating strongly chordal graphs.

2   Definitions

A classical source for all the definitions here is [6]. Let $G = (V, E)$ be a graph with n vertices and m edges. If $S \subseteq V$, then $G[S]$ denotes the induced graph on the vertices of S. Let $N(v)$ denote the neighborhood of a vertex v of G. The closed neighborhood of a vertex v, denoted by $N[v]$, is the set $N(v) \cup \{v\}$. A vertex v in G is simplicial if the induced graph $G([N(v)])$ is complete. We have an analogous characterization for a strongly chordal graph G. A vertex v of G is said to be simple if the sets in $\{N[u] : u \in N[v]\}$ can be linearly ordered by inclusion. An alternate definition is this: vertices u and v of G are said to be compatible if $N[u] \subseteq N[v]$ or $N[v] \subseteq N[u]$ [3]. Then a vertex v is simple if the vertices in $N[v]$ are pairwise compatible. None of the vertices of the graph in Fig. 1a is simple. Thus v_2 is not simple as the vertices v_1 and v_3 in $N[v_2] = \{v_1, v_2, v_3\}$ are not pairwise compatible.

An ordering $v_1, v_2, \ldots, v_n$ of the vertices of G is said to be a perfect elimination ordering, if for each v_i, the induced graph $G([N(v_i)])$ is complete in G. A graph G is chordal if and only if there exists a perfect elimination ordering of its vertices. The vertex order $v_2, v_6, v_4, v_1\ v_3, v_5$ is a perfect elimination ordering of the vertices of the chordal graphs shown in Fig. 1.

A strong elimination ordering of a graph $G = (V, E)$ is an ordering $v_1, v_2, \ldots, v_n$ of V such that the following condition holds: for each i, j, k, and l, if $i < j$, $k < l$ and $v_k, v_l \in N[v_i]$, and $v_k \in N[v_j]$, then $v_l \in N[v_j]$ [3]. Thus a graph G is strongly chordal if it admits a strong elimination ordering, which is a generalization of the notion of perfect elimination ordering used to define chordal graphs. Since the graph shown in Fig. 1a is not strongly chordal, it has no strong elimination ordering. On the other hand, $v_2, v_6, v_4, v_1\ v_3, v_5$ is a strong elimination ordering of the vertices of the graph shown in Fig. 1b.

A graph G is chordal if it has no induced chordless cycle of size greater than 3. A characterization of strongly chordal graphs that is intuitive and analogous to this definition of a chordal graph is based on the notion of a *strong chord*. A strong chord partitions the boundary of an even cycle of size 6 or more into two odd length paths. A graph $G = (V, E)$, is strongly chordal iff it is chordal and every even cycle of size 6 or more has a strong chord. Our second generation method makes use of this.

The strong chord characterization also helps us distinguish Fig. 1a from Fig. 1b. The graph in Fig. 1a is chordal but not strongly chordal because it has no strong chord (known as the Hajos graph in the literature). On the other hand, the graph in Fig. 1b is strongly chordal, since $\{v_3, v_6\}$ (shown as a dashed line segment) is a strong chord.

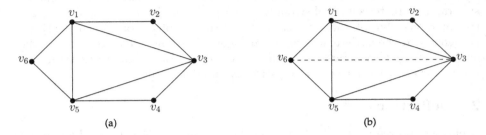

Fig. 1. (a) Chordal but not strongly chordal graph; (b) Strongly chordal graph

3   First Algorithm

Overview: We first generate a chordal graph on n vertices and m edges by using an existing algorithm [8,11]. Next, a perfect elimination ordering (*peo*, for short) for this graph is computed using the lexicographic breadth-first search (Lex-BFS) algorithm, proposed by Rose et al. in [13]. We exploit this *peo* to generate a strongly chordal graph, and simultaneously reduce the *peo* of the input chordal graph into a strong elimination ordering of the resulting strongly chordal graph.

3.1   Details

The neighborhood matrix, $M(G)$, of a graph G is an $n \times n$ matrix whose rows and columns are labeled by the vertices $v_1, v_2, \ldots, v_n$ and its (i,j)-th entry is 1 if $v_i \in N[v_j]$ and 0 otherwise. Note that this matrix has all 1's along the main diagonal. The ordering of the vertices $v_1, v_2, \ldots, v_n$ of G is a strong elimination ordering if and only if the matrix

$$\Delta = \begin{bmatrix} 1 & 1 \\ 1 & 0 \end{bmatrix}$$

is not a submatrix of $M(G)$.

Observation 1 and Definition 2 from Farber [3] are key to our generation algorithm.

Observation 1 [3]. *The row (and column) labels of $M(G)$ correspond to a strong elimination ordering if and only if the matrix M does not contain Δ as a submatrix.*

Definition 1 [3]. *A $(0,1)$ matrix is said to be totally balanced if it does not contain as a submatrix the (edge-vertex) incidence matrix of a cycle of length at least three.*

A Δ-free $M(G)$ is totally balanced and consequently, from Theorem 1, G is strongly chordal.

Theorem 1 [3]. *A graph G is strongly chordal if and only if $M(G)$ is totally balanced.*

Assume that the variable i, $1 \leq i \leq n$, indexes the rows and j, $1 \leq j \leq n$, the columns of $M(G)$. Algorithm 1 is designed to make $M(G)$ Δ-free. For brevity, we use M in place of $M(G)$.

3.2 An Example

Figure 3a shows an example of a chordal graph, G, while Fig. 3b shows the strongly chordal graph generated by Algorithm 1 from it. G is not strongly chordal as the 6-cycles $C_1 = \langle v_0, v_3, v_1, v_4, v_5, v_6, v_0 \rangle$ and $C_2 = \langle v_2, v_6, v_5, v_4, v_1, v_3, v_2 \rangle$ have no strong chords (for example, the edges $\{v_0, v_4\}$ and $\{v_2, v_4\}$ in the respective 6-cycles).

Two occurrences of the $\begin{bmatrix} 1 & 1 \\ 1 & 0 \end{bmatrix}$ submatrix can be detected in M. Following Algorithm 1, by resetting 0 to 1 in each of these occurrences, M becomes Δ-free. This adds two new edges $\{v_0, v_4\}$ and $\{v_2, v_4\}$ to G, which, as we noted earlier, are strong chords of the 6-cycles C_1 and C_2. The resulting graph is now strongly chordal and the perfect elimination ordering v_1, v_0, v_2, v_3 v_4, v_5, v_6 is now a strong elimination ordering of the modified graph.

Algorithm 1. DeltaFreeM

1: Make a list L of index pairs (i, j) in row-major order, such that $i \geq 2$, $j \geq 2$ and $M[i, j] = 0$. If L is empty, STOP.
2: Pick the first pair (i, j) in L, and set $L = L - \{(i, j)\}$.
3: Search the upper left quadrant of M relative to (i, j) for a pair (k, l) such that $M[k, l] = 1$, $M[i, l] = 1$ & $M[k, j] = 1$ (see Fig. 2). Break out of the search loop if such a pair is found and go to Step 4, else go to Step 5.
4: Set $M[i, j] = 1$ and $M[j, i] = 1$. Add the edge $\{v_i, v_j\}$ to G.
5: If L is not empty go to Step 2, else STOP.

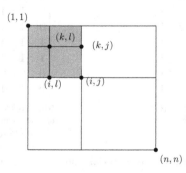

Fig. 2. Search quadrant relative to (i, j), shown shaded

Figure 4a shows another example of a chordal graph with a perfect elimination ordering: v_5, v_1, v_3, v_0 v_2, v_4, v_6, v_7. The graph also happens to be strongly chordal as there are no $\left[\begin{smallmatrix}1&1\\1&0\end{smallmatrix}\right]$ submatrices in its neighborhood matrix, $M(G)$.

Theorem 2. *Algorithm 1 generates a strongly chordal graph, along with a strong elimination ordering.*

Proof: When the neighborhood matrix $M(G)$ of a chordal graph G is input to the Algorithm 1, it returns a Δ-free matrix, where

$$\Delta = \begin{bmatrix}1 & 1\\1 & 0\end{bmatrix}$$

By Definition 2 and Theorem 1, the resulting matrix is totally balanced and hence is now the neighborhood matrix of a strongly chordal graph G'.

Again by Observation 1, we know that the ordering of the vertices $v_1, v_2, \ldots, v_n$ of the graph G' is a strong elimination ordering if and only if Δ is not a submatrix of the neighborhood matrix, $M(G')$. As observed earlier, Algorithm 1 makes sure that no such submatrices are present in $M(G')$. Thus the resulting graph, G', is strongly chordal. ∎

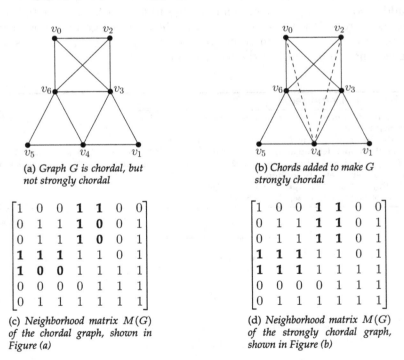

(a) *Graph G is chordal, but not strongly chordal*

(b) *Chords added to make G strongly chordal*

$$\begin{bmatrix}1 & 0 & 0 & \mathbf{1} & \mathbf{1} & 0 & 0\\0 & 1 & 1 & \mathbf{1} & \mathbf{0} & 0 & 1\\0 & 1 & 1 & \mathbf{1} & \mathbf{0} & 0 & 1\\\mathbf{1} & \mathbf{1} & \mathbf{1} & 1 & 1 & 0 & 1\\\mathbf{1} & \mathbf{0} & \mathbf{0} & 1 & 1 & 1 & 1\\0 & 0 & 0 & 0 & 1 & 1 & 1\\0 & 1 & 1 & 1 & 1 & 1 & 1\end{bmatrix}$$

(c) *Neighborhood matrix $M(G)$ of the chordal graph, shown in Figure (a)*

$$\begin{bmatrix}1 & 0 & 0 & \mathbf{1} & \mathbf{1} & 0 & 0\\0 & 1 & 1 & \mathbf{1} & \mathbf{1} & 0 & 1\\0 & 1 & 1 & \mathbf{1} & \mathbf{1} & 0 & 1\\\mathbf{1} & \mathbf{1} & \mathbf{1} & 1 & 1 & 0 & 1\\\mathbf{1} & \mathbf{1} & \mathbf{1} & 1 & 1 & 1 & 1\\0 & 0 & 0 & 0 & 1 & 1 & 1\\0 & 1 & 1 & 1 & 1 & 1 & 1\end{bmatrix}$$

(d) *Neighborhood matrix $M(G)$ of the strongly chordal graph, shown in Figure (b)*

Fig. 3. A chordal graph, a strongly chordal one and their respective neighborhood matrices

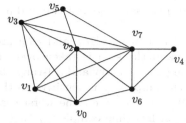

(a) A chordal graph that is also strongly chordal.

$$\begin{bmatrix} 1 & 0 & 1 & 0 & 1 & 0 & 0 & 1 \\ 0 & 1 & 1 & 1 & 1 & 0 & 0 & 1 \\ 1 & 1 & 1 & 1 & 1 & 0 & 0 & 1 \\ 0 & 1 & 1 & 1 & 1 & 0 & 1 & 1 \\ 1 & 1 & 1 & 1 & 1 & 0 & 1 & 1 \\ 0 & 0 & 0 & 0 & 0 & 1 & 1 & 1 \\ 0 & 0 & 0 & 1 & 1 & 1 & 1 & 1 \\ 1 & 1 & 1 & 1 & 1 & 1 & 1 & 1 \end{bmatrix}$$

(b) Neighborhood matrix $M(G)$ of the strongly chordal graph, shown in Figure (a)

Fig. 4. A chordal graph that is also strongly chordal and its neighborhood matrix.

3.3   Complexity Considerations

The time-complexity of Lex-BFS is linear in the number of vertices, n, and the number of edges, m. The time-complexity of Algorithm 1 is dominated by $\Sigma_{k=1}^{|L|} i * j$, where $|L|$ is a count of the number of entries for which $M[i,j] = 0$. Clearly, as both $|L|$ and $i * j$ are in $O(n^2)$, $O(n^4)$ is an upper bound on this sum, which is rather high.

Below, we show that by a suitable preprocessing of the matrix data and an input-sensitive analysis of the time-complexity, the search for a Δ-submatrix in M, for a random input, can be made more efficient than the above worst-case analysis shows.

Four positive integer values, v_u, v_d, h_l, h_r are associated with each 0-entry of the matrix, $M(G)$. These record the runs of that entry value, vertically up and down, horizontally left and right (refer to Fig. 5) respectively. We also record for each row and column of M, its range decomposition into 0's and 1's (see Fig. 6 for an example).

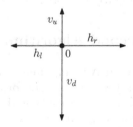

Fig. 5. Runs relative to an entry 0

Let the (i,j)-th entry of $M(G)$ be 0. To check for a Δ-submatrix relative to this entry with a 1 in the upper left quadrant, we search for a 1 in the intersection of 1-range to the left of this entry on the i-th row with a 1-range vertically above this entry on the j-th column (see Fig. 7).

For this we traverse the boundary of the intersected range, looking for a 1 or probe its interior relative to a 0 for a 1 (using the recorded information on horizontal and vertical runs). We terminate the traversal as soon as we find a 1 or report that there is no 1 in this range.

The complexity of this search is: $\Sigma(|R_h| + |R_v|)$, where $|R_h|$ and $|R_v|$ are the sizes of the horizontal and vertical ranges and the sum is taken over all pairs, consisting of a 1-range vertically above the (i,j)-th entry and a 1-range to the left of this entry. This shows that the search-complexity is sensitive to the distribution of 0's and 1's in the matrix $M(G)$ and, except for the worst-case scenario when all the 0 and 1 ranges are of size 1, has time complexity that is of lower order than n^2.

We also update the ranges of 0's and 1's on the i-th and j-th columns and update the neighborhood information of the 0's in v_u, v_d, h_l and h_r, adjoining the (i,j)-th entry. All this work takes linear time.

3.4   Discussion

The proposed algorithm takes n vertices and m edges as input to generate strongly chordal graphs. It is interesting to note that if we start with a a tree as the initial chordal graph we cannot add new edges as a tree is also strongly chordal and thus there is no Δ submatrix present in the neighborhood matrix of a tree. Hence we would get very sparse strongly chordal graphs. Figure 8a shows an example of a tree with a perfect elimination ordering v_3, v_0, v_1 v_2, v_4. The neighborhood matrix is shown in Fig. 8b from which we can see that there is no Δ submatrix present in its $M(G)$. An interesting challenge is to generate strongly a chordal graph from an arbitrary input graph, without generating a chordal graph as an intermediate step.

For testing purposes, at the end of the strongly chordal graph generation process, the recognition Algorithm due to Farber [3] is applied to verify that the graph generated by the Algorithm 1 is strongly chordal. A formal description of the recognition algorithm is given below:

4   Second Algorithm

Overview: A trampoline is a chordal graph G on $2n$ vertices for $n \geq 3$, whose vertex set V is the union of two disjoint sets, $W = \{w_1, w_2, \ldots, w_n\}$ and $U = \{u_1, u_2, \ldots, u_n\}$, such that W is independent and for each i and j, w_i is adjacent to u_j if and only if $i = j$ or $i = j + 1 \mod n$, while $G[U]$ is a complete graph [3]. Figure 9a and Fig. 10a show trampolines on 8 and 10 vertices, respectively.

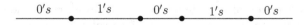

Fig. 6. Range decomposition of a row or column

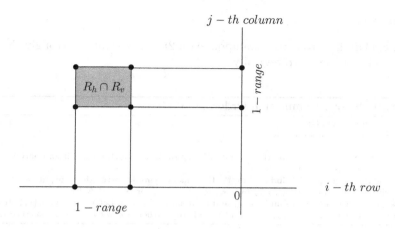

Fig. 7. Searching for a 1 in the intersection of 1-ranges

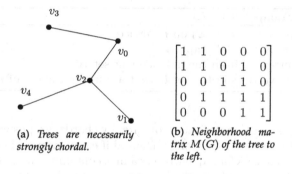

(a) *Trees are necessarily strongly chordal.*

(b) *Neighborhood matrix $M(G)$ of the tree to the left.*

Fig. 8. A tree and its neighborhood matrix.

The algorithm of this section is based on the following forbidden subgraph characterization of strongly chordal graphs by Farber:

Theorem 3 [3]. *A chordal graph is strongly chordal if and only if it contains no induced trampoline.*

Since a trampoline is provably chordal (any permutation of the vertices in W, followed by any permutation of the vertices in U is a perfect elimination ordering), we exploit Theorem 3 to generate a strongly chordal graph from a trampoline and then introduce additional edges, if needed, by applying the completion algorithm. In Sect. 4.2, we extend the scope of this method to generate dense strongly chordal graphs from a network of trampolines.

4.1   Details

We first explain how to turn a trampoline on $2n$ vertices into a strongly chordal graph by adding edges strategically.

Algorithm. Strong Elimination Ordering [3]

Input: A graph $G = (V, E)$
Output: A strong elimination ordering
1: Set $n \leftarrow |V|$.
2: Let $V_0 = V$ and let $(V_0, <_0)$ be the partial ordering on V_0 in which $v <_0 u$ if and only if $v = u$. Let $V_1 = V$, and set $i \leftarrow 1$.
3: Let G_i be the subgraph of G induced by V_i. If G_i has no simple vertex then output G_i and stop. Otherwise, define an ordering on V_i by $v <_i u$ if $v <_{i-1} u$ or $N_i[v] \subset N_i[u]$.
4: Choose a vertex v_i which is simple in G_i and minimal in $(V_i, <_i)$. Let $V_{i+1} = V_i - \{v_i\}$. If $i = n$ then output the ordering $v_1, v_2, \ldots, v_n$ of V and stop. Otherwise, set $i \leftarrow i+1$ and go to step 2.

Algorithm 2. TrampolineToSCG

Input: A trampoline $G = (V, E)$ on n vertices
Output: A strongly chordal graph G
1: Choose an vertex w_i from the independent set W
2: Join w_i to the vertices in the set U that are not neighbors of w_i

---

From an alternate characterization of strongly chordal graphs we know that a chordal graph G is necessarily strongly chordal if every even cycle of length 6 or more has a strong chord. Consequently, G cannot contain an induced trampoline. Therefore, the strategy underlying our algorithm is to introduce strong chords in even cycles of size 6 and greater.

Consider the trampoline shown in Fig. 9a. The even cycle of length 8, namely $\langle w_1, u_1, w_2, u_2, w_3, u_3, w_0, u_0, w_1 \rangle$, has no strong chord. Algorithm 2 adds edges $\{w_1, u_3\}$ and $\{w_1, u_2\}$ by joining the independent vertex w_1 to its non-neighbors u_2 and u_3 in the set U. This turns the trampoline into the strongly chordal graph of Fig. 13a. While $\{u_0, u_2\}$ or $\{u_3, u_1\}$ is a strong chord for every even cycle of length 6, the outer 8-cycle has no strong chord. Joining w_1 and u_3 splits this 8-cycle into a 4-cycle $\langle w_1, u_0, w_0, u_3, w_1 \rangle$ and a 6-cycle $\langle w_1, u_1, w_2, u_2, w_3, u_3, w_1 \rangle$. However, there is no strong chord in the newly created 6-cycle. This is rectified by adding an edge between w_1 and u_2.

Another example is shown in Fig. 10, where a trampoline of size 10 is turned into a strongly chordal graph by adding 3 strong chords.

This approach can be extended to any trampoline of size $n \geq 3$ and ensures a strong chord in every even cycle of length 6 and more. This means adding $n - 2$ additional edges in a trampoline of size $2n$. Below, we provide a formal proof of this claim.

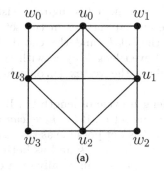
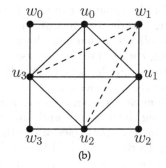

Fig. 9. (a) Trampoline(G) (a) Strongly chordal graph(G')

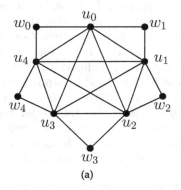
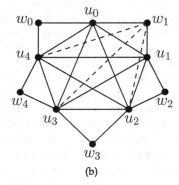

Fig. 10. (a) Trampoline(G) (b) Strongly chordal graph(G')

Proof of Correctness of Algorithm 3: We first show that every cycle of length 6 has a strong chord. Four cases arise. A cycle of length 6 in the trampoline, plus added chords has (1) no vertex; (2) one vertex; (3) two vertices; (4) three vertices from the set W. These are shown in Fig. 11, in which solid disks represent vertices belonging to the set U and the circles represent those belonging to the set W.

Fig. 11. Strong chord in a 6-cycle

In the first case, since the vertices all belong to the set U, there is a chord between every pair of non-adjacent vertices and the chord shown is a strong chord. For the same reason the chords shown in the next two figures exist and are strong chords. In the last case, one of the circled vertices in the set W is joined to all the remaining three vertices in the set U and the chord shown is a strong chord. This is because any induced even length cycle with vertices alternating from the sets U and W must include the vertex w_i on which all added chords are incident.

Consider now the case of any larger even length cycle of length $2k$. Let l of these belong to the set W and the rest $2k - l$ to the set U. If $l = k$ we can argue as in the last case for cycles of length 6. Now, $l > k$ is not possible as two vertices belonging to the set W will have to be adjacent. Thus $l < k$ and as both must be odd or even, $k - l \geq 2$. If we arrange these vertices schematically on a circle, each pair of vertices of W must be separated by at least one vertex of U. Thus there must be one pair of consecutive vertices of W that are separated by at least two vertices of U. Consider such an interval of vertices of U, bounded at each end by a vertex of W. Let us call these vertices in W, x and y with the latter placed anticlockwise with respect to x along the boundary of the circle. Consider the first vertex of U that lies anticlockwise with respect to x (see Fig. 12). We traverse the boundary of the circle counting 4 vertices, inclusive of both the initial and terminal points. If the terminal point is in W, and it happens to be the vertex at which we have extra chords, then this is a strong chord; if not, consider the next vertex in U anticlockwise with respect to the initial point (there exists one since this interval between x and y has at least two points of the set U) and the next point anticlockwise with respect to the terminal point, which has to be in U. There exists a chord joining these two points and this is a strong chord of the even cycle of length $2k$.

We need to add $n - 2$ chords to a selected vertex of W, as there are $n - 2$ even-length cycles of sizes 6 to n in the modified trampoline, incident on the selected vertex, in each of which a vertex of U alternates with a vertex of W. For each of these cycles an added chord incident on the selected vertex of W acts as a strong chord.

For the example of Fig. 10(b), these are:

1. a 6-cycle, $\langle w_1, u_0, w_0, u_4, w_4, u_3, w_1 \rangle$, with $w_1 u_4$ as a strong chord
2. an 8-cycle, $\langle w_1, u_0, w_0, u_4, w_4, u_3, w_3, u_2, w_1 \rangle$, with $w_1 u_3$ as a strong chord and
3. a 10-cycle, $\langle w_1, u_0, w_0, u_4, w_4, u_3, w_3, u_2, w_2, u_1, w_1 \rangle$, with $w_1 u_2$ as a strong chord

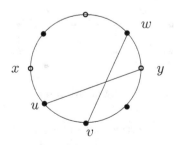

Fig. 12. Strong chord in an even cycle on 8 vertices

Adding More Edges: In [10], Odom showed that strongly chordal graphs, in addition to other graph classes like chordal graphs, constitute a completion class. This allows us to add an edge at a time to a strongly chordal graph to reach the complete graph, remaining in the class all throughout. The completion Algorithm 3 is based on Theorem 4.

Algorithm 3. SCGCompletion [10]

Input: A strongly chordal graph G, a strong elimination ordering θ, and the number of edges m
Output: A strongly chordal graph G
1: $G_0 \leftarrow G$
2: $E_0 \leftarrow E$
3: $s \leftarrow m - m'$ $\triangleright$ m' is the no. of edges in G (before applying the completion algorithm)
4: **for** $i \leftarrow 1$ to s **do**
5:      $k_i \leftarrow \max\{j | \deg(v_j) < n - 1\}$
6:      $m_i \leftarrow \max\{l | v_{k_i}, v_l \notin E_{i-1}\}$
7:      $e_i \leftarrow v_{k_i}, v_{m_i}$
8:      $E_i \leftarrow E_{i-1} \cup \{e_i\}$
9:      $G_i \leftarrow (V, E_i)$
10: **end for**

Theorem 4 [10]. *Let $G = (V, E)$ be a connected graph of order n and size m. Let $G_0 = G$, and define the sequence of graphs $G_0, G_1, \ldots, G_s$ using Algorithm 3. If θ is a strong elimination ordering for G, then θ is a strong elimination ordering for each G_i, $i = 1, 2, \ldots, s$.*

The completion algorithm takes a strongly chordal graph, a strong elimination ordering, and the number of edges (m) as input and produces a strongly chordal graph. We use the recognition algorithm by Farber [3], to generate a strong elimination ordering. Based on the strong elimination ordering, we choose a pair of vertices k_i and m_i according to the conditions mentioned in the algorithm and introduce an edge between them. We iterate s times to add s additional edges to meet the target of m edges.

We illustrate the completion algorithm by means of an example. Consider the strongly chordal graph (G') shown in Fig. 13a where $m' = 16$. Let $m = 18$. We need to add two more edges as $s = 2$, using the completion algorithm. From Farber's recognition algorithm, we obtain the following strong elimination ordering: w_2, w_3, w_0, w_1, u_3, u_2, u_1, u_0. We choose a pair of vertices from the ordering with no edge between them. In the first iteration, we introduce an edge between u_0 and w_3 and in the next iteration, we introduce an edge between u_0 and w_2. This gives a strongly chordal graph G'' for the given n and m.

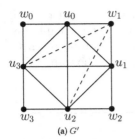 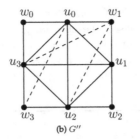 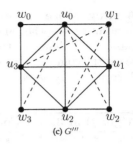

Fig. 13. Strongly chordal graph generation

4.2 Network of Trampolines

To generate a greater variety of strongly chordal graphs, we construct a trampoline network, and apply Algorithm 3 as a subroutine to turn each trampoline of this network into a strongly chordal graph. A network of trampolines and a strongly chordal graph derived from it are shown in Fig. 14.

4.3 Complexity Considerations

To create a trampoline of size $2n$ takes $O(n^2)$ time, this being the size of an adjacency list to represent this graph. The complexity of the completion procedure is in $O(n^3)$, this being the complexity of generating a strong elimination ordering using Farber's algorithm. Thus the time complexity of this method is in $O(n^3)$ for generating a strongly chordal graph from a single trampoline. The time complexity of generating a strongly chordal graph from a trampoline network is in $O(k + \Sigma_{i=1}^{k} n_i^3)$, where k is the number of nodes in the trampoline network and n_i is the size of the i-th trampoline.

5   Third Algorithm

Preamble: Farber's intersection graph characterization of strongly chordal graphs [2] is analogous to a similar characterization of chordal graphs by Gavril [5]. Şeker et al. [14] exploited this characterization of chordal graphs to obtain an algorithm for generating them. In this section, we propose an algorithm to show that strongly chordal graphs can also be generated, using their intersection graph characterization.

The following essential definitions from Farber [2] underlie this characterization. Let r be the root T of an edge-weighted tree. The edge-weights are positive numbers that can be conveniently interpreted as the lengths of the edges.

Definition 2. *The weighted distance from a node u to a node v in T, denoted by $d_T^*(u, v)$, is the sum of the lengths of the edges of the (unique) path from u to v.*

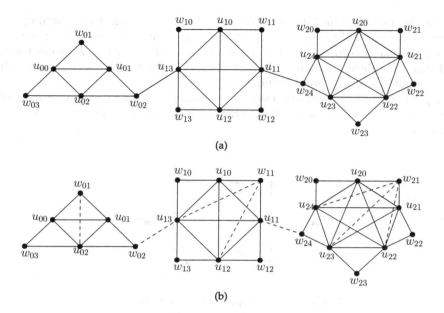

Fig. 14. (a) Network of Trampoline(G) (b) Strongly chordal graph(G')

Definition 3. *Let T_1 and T_2 be two subtrees of T. Subtree T_1 is full with respect to T_2, denoted by $T_1 > T_2$, if for any two vertices $u, v \in T_2$ such that $d_T^*(r, u) \leq d_T^*(r, v)$, $v \in V(T_1)$ implies that $u \in V(T_1)$.*

Definition 4. *A collection of subtrees $\{T_1, T_2, \ldots, T_n\}$ of T is compatible if for each pair of subtrees T_i and T_j either $T_i > T_j$ or $T_j > T_i$.*

Using the definitions above, Farber established the following intersection graph characterization for strongly chordal graphs.

Theorem 5 [2]. *A graph is strongly chordal if and only if it is the intersection graph of a compatible collection of subtrees of a rooted, weighted tree, T.*

5.1   The Algorithm

Let M be an adjacency matrix whose rows correspond to a compatible collection of subtrees, $\{T_1, T_2, \ldots, T_n\}$ of a rooted, weighted tree T on n vertices as in Theorem 5 and columns correspond to the vertices $\{w_1, w_2, \ldots, w_n\}$ of T, arranged from left to right in order of increasing distance from the root, w_n.

Our main observation is that Definition 3 can be re-interpreted to imply that the matrix M cannot have $\Delta = \left[\begin{smallmatrix} 1 & 1 \\ 0 & 1 \end{smallmatrix}\right]$ as a sub-matrix. More precisely, if i and j are two rows of M, corresponding to compatible subtrees T_i and T_j of T, then there cannot exist columns k and l that intersect these two rows to create Δ. Thus M belongs to the class of 0-1 matrices that do not have Δ as a submatrix.

We get more insight into the structure of such a matrix, by examining the constructive proof of the necessity part Theorem 5, cited above. Starting with a strong elimination ordering of the vertices of G, the edges of a tree T_G are defined, and then the subtrees rooted at the vertices of this tree whose intersection graph is the input strongly chordal graph.

By simulating this construction on the example strongly chordal graph of Fig. 15 for the strong elimination order: $\langle v_6, v_4, v_5, v_1, v_2, v_3 \rangle = \langle w_1, w_2, w_3, w_4, w_5, w_6 \rangle$,

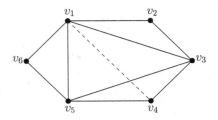

Fig. 15. Hajós graph with a strong chord

we obtain the subtree-vertex incidence matrix of Table 1.

Table 1. Tree-vertex incidence matrix for the Hajós graph plus a strong chord

	w_6	w_5	w_4	w_3	w_2	w_1
T_1	0	0	0	0	0	1
T_2	0	0	0	0	1	0
T_3	0	0	0	1	1	1
T_4	0	0	1	1	1	1
T_5	0	1	1	0	0	0
T_6	1	1	1	1	1	0

The above matrix does not have Δ as a submatrix. It also has an upper triangle of 0's only, bordered with an antidiagonal of 1's. On each row, corresponding to a T_i, there is a run of 1's starting from the 1 on the antidiagonal and going right, followed by a run of 0's. Figure 16 shows the strongly chordal graph based on the above matrix. It is seen to be isomorphic to the Hajós graph with an added chord, shown in Fig. 15.

This suggests an algorithm for generating strongly chordal graphs. Generate an $n \times n$ matrix of 0's and 1's that have the properties described above. The strongly chordal graph corresponding to this matrix has a vertex, v_{T_i} for the

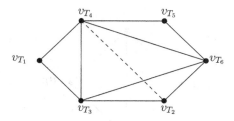

Fig. 16. Hajós graph with a strong chord from matrix

Table 2. Tree-vertex incidence matrix for a 4-cycle with a chord

	w_4	w_3	w_2	w_1
T_1	0	0	0	1
T_2	0	0	1	0
T_3	0	1	1	1
T_4	1	1	1	1

row T_i, and an edge joins two vertices if the corresponding rows have a common vertex.

Here are some more examples. From the tree-vertex incidence matrix of Table 2, we can generate the strongly chordal graph of Fig. 17.

The graph that can be generated from the following tree-vertex incidence matrix of Table 3 is shown in Fig. 19. The isomorphism of this graph with the graph of Fig. 13(a) is obvious. The trees corresponding to the rows are subtrees of the weighted tree of Fig. 19. The edge-weights are set so that the distance from the root, w_8, of the tree to a node w_i is $8 - i$ (Fig. 18).

Table 3. Tree-vertex incidence matrix for Trampoline Plus

	w_8	w_7	w_6	w_5	w_4	w_3	w_2	w_1
T_1	0	0	0	0	0	0	0	1
T_2	0	0	0	0	0	0	1	0
T_3	0	0	0	0	0	1	0	0
T_4	0	0	0	0	1	1	0	0
T_5	0	0	0	1	1	0	0	0
T_6	0	0	1	1	1	0	0	1
T_7	0	1	1	1	1	0	1	1
T_8	1	1	1	1	1	1	1	0

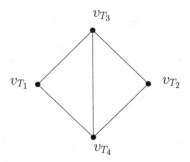

Fig. 17. A 4-cycle with a diagonal

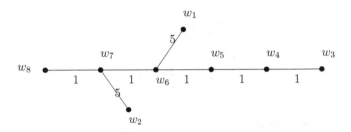

Fig. 18. Weighted tree corresponding to the tree-vertex incidence matrix of Table 3

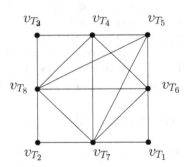

Fig. 19. A strongly chordal graph from a trampoline

While the matrix of Table 3 satisfies the necessary condition of not having Δ as a submatrix, its structure is a little different from the matrices we generated earlier. The rows labeled T_6 and T_7 do not have a run of 1's starting with 1 on the antidiagonal, followed by all 0's. The run of 0's is interrupted by the appearance of a run of 1's at the end. No matter, in all cases the trees that the rows represent are pairwise compatible. We thus take up the problem of generating a 0-1 matrix without Δ as a submatrix and use this to generate strongly chordal graphs.

5.2 Generating a Δ-Free Submatrix

Let M be an $n \times n$ matrix of 0's and 1's whose entries are all set to 0. Next,

- all antidiagonal entries are set to 1;
- each row i is reset to have a run of 1's of random length (not exceeding $n - j$, where j is the column index of the antidiagonal entry), that runs right from the 1 on the antidiagonal;

Such a 0-1 matrix provably does not have Δ as a submatrix (see Table 1 for an example). This is because the only 1's that have 0's to their left in the same row are the 1's on the antidiagonal. However any row above such a row intersects the columns that contain such a 0-1 pair in 0's, pre-empting the formation of a Δ submatrix.

We now select a random entry (i, j), such that $i + j > n + 1$. If $M[i, j]$ is 1, we select another (i, j). Otherwise, we tentatively reset this 0 to a 1 and check the second, third and fourth quadrants, relative to (i, j) (see Fig. 20) if this creates a Δ submatrix. If it does, we restore the entry to 0 again and choose another random entry. This is repeated until we have exhausted the preset (or user-prescribed) number of random trials. The resulting matrix M remains Δ-free.

A cursory analysis indicates that we might need to examine $O(n^2)$ entries to check if M remains Δ-free when the (i, j)-th entry is changed from 0 to 1. However, preprocessing M for range-searching as suggested in the analysis of the first algorithm makes the search more efficient for an average input. It is to be noted however that such an improvement is sensitive to the distribution of 0's and 1's in the matrix M.

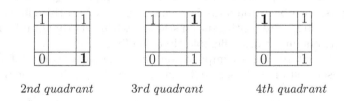

2nd quadrant   3rd quadrant   4th quadrant

Fig. 20. Search quadrants for a Δ submatrix, relative to the 1 in bold

The time required for checking for subtree intersection is in $O(n^2)$ as for each subtree we check the entries below the entry 1 on the antidiagonal for entries that are 1. Each such entry indicates an intersection with the subtree corresponding to the row in which this 1 appears.

Thus the overall complexity of this method is in $O(n^2)$ if take advantage of preprocessing the matrix M to execute the first part and the number of 0's we attempt to change into a 1 is in $O(n)$.

6   Experimental Results

As an aid to experimentation and visualization, we have implemented all the proposed algorithms in Python. The software was developed in Python 2.7.17 running under Ubuntu 18.04.4 LTS operating system on a slate with a 64-bit Intel Core i5 CPU U 470 1.33 GHz × 4 processor and 3.6 GiB RAM. The source codes for the implementations of all the three algorithms are available at https:// github.com/zamiljitu/SCGGeneration.

In an appendix, we have shown the outputs from the three different algorithms for $n = 10, 14, 20$ and 26, where n is the number of vertices in the strongly chordal graphs.

The legends at the top of these graph-plots, also indicate the number of edges. The number of edges is either user-specified or are determined by the algorithm. For the first algorithm, these are user-specified; the same goes for the trampoline-based generation method, if we choose the completion-based version. In this case, the target number of edges are reached by Odom's completion algorithm for strongly chordal graphs.

The second set of 4 figures show the output of a variation of the second (trampoline-based) method where a minimal number of edges are added to a trampoline to make it strongly chordal. The fourth set of graph-plots are from the third method, where the number of edges added are determined by the algorithm.

7   Conclusions

In this paper, we have presented three novel algorithms for generating strongly chordal graphs, each based on a distinct characterization of this class of graphs. The first two methods are contingent on starting with a chordal graph. The third method is not, making it more flexible than the first two.

It would be interesting to improve on the time-complexities of the three algorithms discussed or find more efficient ways of generating strongly chordal graphs.

A challenging problem is to come up with an algorithm for generating strongly chordal graphs uniformly at random. This requires counting the number of labeled strongly chordal graphs on n vertices and m edges.

As an aid to experimentation and visualization, we have implemented all the proposed algorithms in Python.

Appendix

See Figs. 21, 22, 23 and 24.

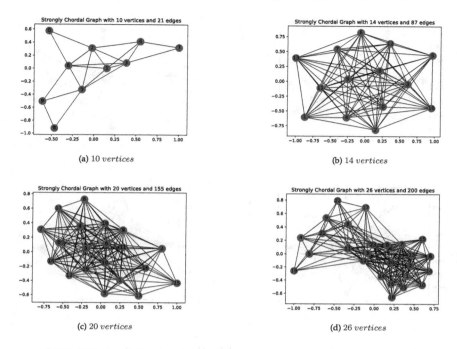

Fig. 21. Strongly chordal graphs generated by the first method

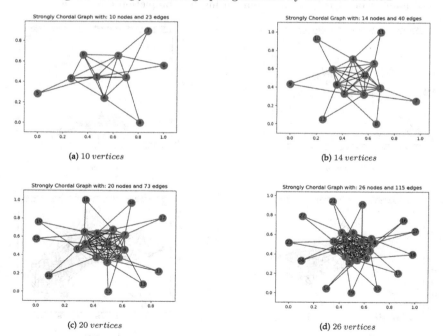

Fig. 22. Strongly chordal graphs generated by the second method that adds a minimal number of edges to a trampoline

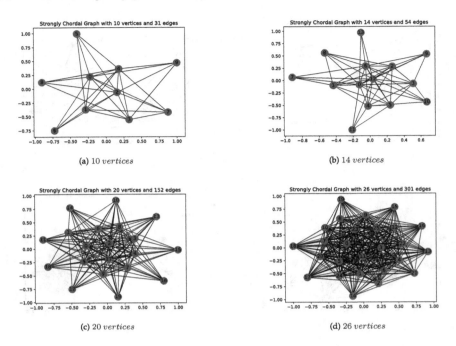

Fig. 23. Strongly chordal graphs generated by the second method that adds a user-specified number of edges to a trampoline using the completion procedure

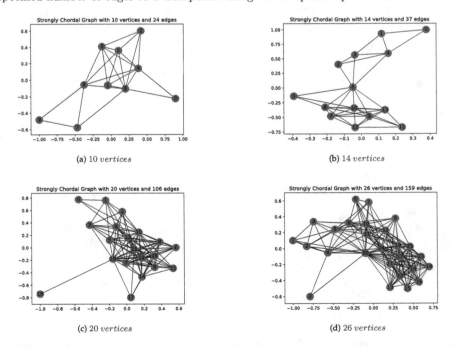

Fig. 24. Strongly chordal graphs generated by the third method

References

1. Dahlhaus, E., Manuel, P.D., Miller, M.: A characterization of strongly chordal graphs. Discret. Math. **187**(1–3), 269–271 (1998)
2. Farber, M.: Applications of 1.p. duality to problems involving independence and domination. Ph.D. thesis, Rutgers University (1982)
3. Farber, M.: Characterizations of strongly chordal graphs. Discret. Math. **43**(2–3), 173–189 (1983)
4. Farber, M.: Domination, independent domination, and duality in strongly chordal graphs. Discret. Appl. Math. **7**(2), 115–130 (1984)
5. Gavril, F.: The intersection graphs of subtrees in trees are exactly the chordal graphs. J. Comb. Theory Ser. B **16**(1), 47–56 (1974)
6. Golumbic, M.C.: Algorithmic Graph Theory and Perfect Graphs. Annals of Discrete Mathematics, vol. 57. North-Holland Publishing Co., Amsterdam (2004)
7. Liao, C., Chang, G.J.: k-tuple domination in graphs. Inf. Process. Lett. **87**(1), 45–50 (2003)
8. Markenzon, L., Vernet, O., Araujo, L.H.: Two methods for the generation of chordal graphs. Ann. OR **157**(1), 47–60 (2008). https://doi.org/10.1007/s10479-007-0190-4
9. McKee, T.A.: A new characterization of strongly chordal graphs. Discret. Math. **205**(1–3), 245–247 (1999)
10. Odom, R.M.: Edge completion sequences for classes of chordal graphs. Master's thesis, Naval Postgraduate School (1995)
11. Rahman, M.Z.: Chordal Graphs and their Relatives: Algorithms and Applications. Ph.D. thesis, University of Windsor (2020)
12. Rahman, M.Z., Mukhopadhyay, A., Aneja, Y.P.: A separator-based method for generating weakly chordal graphs. Discret. Math. Algorithms Appl. **12**(4), 2050039:1–2050039:16 (2020)
13. Rose, D.J., Tarjan, R.E., Lueker, G.S.: Algorithmic aspects of vertex elimination on graphs. SIAM J. Comput. **5**(2), 266–283 (1976)
14. Seker, O., Heggernes, P., Ekim, T., Taskin, Z.C.: Generation of random chordal graphs using subtrees of a tree. CoRR, abs/1810.13326 (2018)

A New Bio-heuristic Hybrid Optimization for Constrained Continuous Problems

Prathap Siddavaatam and Reza Sedaghat[✉]

OPRA-Labs, Ryerson University,
350 Victoria Street, Toronto, ON M5B 2K3, Canada
{prathap.siddavaatam,rsedagha}@ee.ryerson.ca
https://www.ee.ryerson.ca/opr/

Abstract. A novel bio-inspired evolutionary algorithm known as MoFAL is presented in this article. The proposed algorithm (MoFAL) is based on the hybrid amalgamation of two nature inspired methods based on Moth Flame Optimization and Ant Lion Optimizer algorithms. It is well known that elitism forms an important characteristic of evolutionary algorithms that allows them to maintain the best fitness(es) obtained at any stage of the optimization process. MoFal is bench-marked using a set of 23 classical benchmark functions employed to test different characteristics during its evolutionary computation process. Numerical experiments demonstrate that the solutions of the constrained optimization problems like Pressure Vessel and the Rolling Element Bearing designs found using our algorithm are highly accurate and their convergence is comparatively fast coupled with improved exploration, local optima avoidance and exploitation. The results clearly exhibit that MoFAL algorithm is capable of finding superior optimal designs for our case study problems that include diverse search spaces. Our algorithm is able to determine global solutions of constrained optimization problems more efficiently than traditional evolutionary algorithms, and also avoid the occurrence of premature phenomena during its convergence process.

Keywords: Ant Lion Optimizer · Moth Flame Optimization algorithm · Algorithm design and analysis · Classification algorithms · Mathematical model · Heuristic algorithms · Constrained Optimizations

1 Introduction

Optimization model is a mathematical model that refers to picking the optimal solution from all feasible solutions. Searching for the best solution for a given optimization problem is difficult due to the constraints and complexity of the problem. In the recent years, evolutionary optimization algorithms have been proposed to solve various optimization problems and real world engineering design problems Kennedy (1995).

© Springer-Verlag GmbH Germany, part of Springer Nature 2021
M. L. Gavrilova and C. J. K. Tan (Eds.): Trans. on Comput. Sci. XXXVIII, LNCS 12620, pp. 76–97, 2021.
https://doi.org/10.1007/978-3-662-63170-6_5

A deterministic algorithm arrives at similar results for a given problem with identical initial starting point as these algorithms often get entrapped in a local minima and fails to arrive at global minima. Since real world problems have in-numerous local solutions, deterministic algorithms are futile and unreliable in find true global optimum. Randomness is the main trait of stochastic algorithms. They employ random operators in order to avoid local minimums and stochastic operators which enable algorithms to obtain different solutions for a given problem in every run Black (2005).

Meta-heuristic algorithms search for the global optimum in a search space by creating one or more random solutions for a given problem Holland (1992). Hence, these algorithms have following advantages: problem independence, evolution independence, local minima evasion and natural optimization inspirations makes these algorithms makes it simple and follow a general and common framework, which imparts us scope to improve these algorithms with hybridization. Consequently in this article, we utilize the best of the characteristics of most popular and efficient evolutionary algorithms as candidates for hybridization: Moth Flame Optimization Mirjalili (2015b) and Ant Lion Optimizer algorithms Mirjalili (2015a).

Optimization exists in almost every aspect right from engineering design, business planning, computer networking and to even leisure travel. Our ultimate goal is to optimize something such as quality, profit and time. In optimization of a design, the design objective could be simply to minimize the cost of production or to maximize the efficiency of production.

"Hybrid algorithm" Hybrid Algorithms -Evolutionary (2020) does not necessarily refer to simply combining multiple algorithms to solve a different problem – many algorithms can be considered as combinations of simpler pieces – but only to combining algorithms that solve the same problem, but differ in other characteristics, notably performance.

The article is organized as follows: Sect. 1 outlined the introduction. Section 2 discusses literature review. Section 3 deals with the concept of hybridization and implementation of our. Section 4 showcases bench-marking and results of 23 fitness functions used. Section 5 analyzes algorithm performance in two real world engineering design problem and Sect. 6 concludes the paper and discusses possible future scope and improvements.

2 Related Works

Stochastic algorithms are categorized as Heuristic and Meta-heuristics algorithms. In heuristic algorithms, quality solutions to a complex optimization problem can be achieved in a reasonable amount of time, but there is no guarantee that optimal solutions are obtained. These algorithms have a good success rate albeit in some specific instances Stojanovic (2017). Meta-heuristic optimization algorithms perform comparatively better than heuristic algorithms and have proven their efficiency for solving optimization problems using iterations and stochastic behaviour Stojanovic (2017). They produce optimal solutions within a reasonable and comparatively better time frame. The complexity of the problem of interest makes it impossible to search every possible solution or combination and the goal is to find acceptable optimal solution within specified time frame.

A hybrid algorithm is an algorithm that combines two or more other algorithms that solve the same problem, either choosing one (depending on the data), or switching between them over the course of the algorithm. This is generally done to combine desired features of each, so that the overall algorithm is better than the individual components. "Hybrid algorithm" does not refer to simply combining multiple algorithms to solve a different problem – many algorithms can be considered as combinations of simpler pieces – but only to combining algorithms that solve the same problem, but differ in other characteristics, notably performance.

There is a recent trend in formulating hybrids of efficient evolutionary algorithms to further enhance the performance of these algorithms and expand its applicability across multiple domains Grosan and Abraham (2007). In Sayed and Hassanien (2018) a hybrid algorithm based on using moth-flame optimization (MFO) algorithm with simulated annealing (SA), namely SA-MFO is developed taking the ability to escape from local optima mechanism of SA and fast searching and learning mechanism for guiding the generation of candidate solutions of MFO. Lévy-flight moth-flame optimization (LMFO) algorithm was proposed in Li et al. (2016) Li, Zhou, Zhang, and Song to improve the performance of MFO. The proposed LMFO used Lévy-flight strategy in the searching mechanism of MFO. A Time-constrained Genetic-Moth-Flame Optimization (TG-MFO) is developed and applied it for efficient energy optimization in smart homes and buildings in paper Ullah and Hussain (2019). A hybrid meta-heuristic optimization algorithm Particle Swarm Optimization-Moth Flame Optimizer (HPSO-MFO) method is proposed in Jangir et al. (2016) Jangir, Parmar, Jangir, Kumar, Trivedi, and Bhoye by using a combination of PSO used for exploitation phase and MFO for exploration phase in an uncertain environment.

In Parvathi and Rajendran (2016), proposes a new image segmentation approach based on Fuzzy C Means (FCM) and Ant Lion Optimization (ALO). FCM has the ability to represent ambiguous information in a more robust way. Bioinspired algorithms such as ALO have the ability to find optimal parameters in search spaces. These characteristics of FCM and ALO have been utilized in the paper for improving image segmentation. A hybrid Ant lion optimization (ALO) and Ant colony optimization (ACO), ACO–ALO algorithm was proposed to solve the data clustering problem in Mageshkumar et al. (2018) Mageshkumar, Karthik, and Arunachalam. Additionally Cauchy's mutation operator is added with this proposed algorithm to avoid the local minima trapping problem. Both artificial neural networks and the ant-lion optimizer, which is a recent intelligent optimization technique, were employed to comprehend the related system and perform some prediction applications over electroencephalogram time series in Köse (2018). A Kmean with Antlion optimisation algorithm is realized in Majhi and Biswal (2018). In this ALO, randomly selection of node(ant) and ant lion is done in the initialization. In addition, calculation of Euclidean distance is done between each pair of sensor nodes, and then the fitness value of ant and ant lion is calculated. Finally, the ant lion fitness value sorted to find the best ant lion is named as elite ant lion.

In Xu et al. (2017) Xu, Liu, and Su an improved GWO algorithm combined with Cuckoo Search (CS) is proposed. By introducing the global-search ability of CS into GWO to update its best three solutions that are alpha-wolf, beta-wolf and delta-wolf, the search ability of GWO is strengthened, and the local minima shortcoming of GWO is offset. A Cuckoo Search (CS) and Firefly algorithm (FA) hybrid CS-FA technique is introduced in Elkhechafi et al. (2018) Elkhechafi, Hachimi, and Elkettani, which runs CS and FA in parallel and update the population and search for global optimum. By making use of the advantages of Cuckoo Search (CS) algorithm and Particle Swarm Optimization (PSO), a hybrid optimization algorithm of PSO and CS was proposed in Wang (2011). By CS-PSO, the search area of PSO was extended, and the defect of PSO is that it falls easily into point of local minimum that was improved. An effective hybrid cuckoo search and genetic algorithm (HCSGA) is introduced in Ganesan et al. (2014) Ganesan, S.G., N, and Janardhanan for solving engineering design optimization problems.

A particle swarm optimization (PSO) and firefly algorithm (FA) hybrid is modeled in Xia et al. (2017) Xia, Gui, He, Xie, Wei, Xing, Wu, and Tang which divides the whole population into two sub-populations selecting FA and PSO as their basic algorithm to carry out the optimization process. A GA-FA-PS algorithm is introduced in paper Wahid et al. (2018) Wahid, Ghazali, and Shah, in which genetic algorithm (GA) has been applied to generate the initial solution for balancing the exploration and exploitation at the initial stage. In the second stage, crossed over operator is embedded in firefly changing position to improve local search which ultimately enhances local convergence. To further improve the local and global convergence rate, pattern search (PS) is introduced which is used to obtain the most optimal solution or at least the solution better than the solution provided by the standard firefly algorithm. An effective combination of two metaheuristic algorithms, namely Firefly algorithm and the Differential evolution, has been proposed in Sarbazfard and Jafarian (2016). In this algorithm, The DE algorithm performs the mutation and crossover on the one firefly which couldn't find a brighter firefly in FFA.

3   Background

3.1   Moth Flame Optimization Algorithm

Moths are fancy insects, which are highly similar to the family of butterflies. Basically, there are over 160,000 various species of this insect in nature. They have two main milestones in their lifetime: larvae and adult. The larvae is converted to moth in cocoons. The most interesting fact about moths is their special navigation methods in night. They have been evolved to fly in night using the moon light. They utilized a mechanism called transverse orientation for navigation. In this method, a moth flies by maintaining a fixed angle with respect to the moon, a very effective mechanism for travelling long distances in a straight path Frank (2006). Since the distance to moon is insurmountable for the moth, this mechanism guarantees flying in straight line.

A similar navigation method can be implemented by humans too. For example, suppose that the moon is in the south side of the sky and a human wants to go the east. Then, if he keeps moon of his left side when walking, he would be able to move towards the east on a straight line. Despite the effectiveness of transverse orientation, we usually observe that moths fly spirally around the lights. In fact, moths are tricked by artificial lights and show such behavior. This is due to the inefficiency of the transverse orientation, in which it is only helpful for moving in straight line when the light source is very far. When moths see a human-made artificial light, they try to maintain a similar angle with the light to fly in straight line. Since such a light is extremely close compared to the moon, however, maintaining a similar angle to the light source causes a useless or deadly spiral fly path for moths Frank (2006). It may be observed in that the moth eventually converges towards the light. This type of behaviour is modeled mathematically in Moth-Flame Optimization (MFO) algorithm Mirjalili (2015b).

MFO is inspired by the navigation method of moths in nature called transverse orientation. In this method, a moth flies by maintaining a fixed angle with respect to the moon; a very productive technique in travelling long distance especially in a straight path Gaston et al. (2013) Gaston, Bennie, Davies, and Hopkins. As the moon lies at very large distance from the moth fly, this mechanism guarantees flying in a straight path. However, this straight path travelling is hugely affected by man-made artificial light. Since these artificial lights are so close to moths; they are tricked and often end up in flying spirally around these lights.

3.2 Ant Lion Optimization Algorithm

The Ant Lion Optimizer, known as ALO or Antlion Optimize, is a meta-heuristic Mirjalili (2015a) algorithm that models the interaction of ants and ant-lions in nature. Antlions (doodlebugs) belong to the Myrmeleontidae family and Neuroptera order (net-winged insects). The lifecycle of antlions includes two main phases: larvae and adult. A natural total lifespan can take up to 3 years, which mostly occurs in larvae (only 3–5 weeks for adulthood). Antlions undergo metamorphosis in a cocoon to become adult. They mostly hunt in larvae and the adulthood period is for reproduction.

Their names originate from their unique hunting behaviour and their favourite prey. An antlion larvae digs a cone-shaped pit in sand by moving along a circular path and throwing out sands with its massive jaw Scharf et al. (2008) Scharf, Subach, and Ovadia, Scharf and Ovadia (2006). After digging the trap, the larvae hides underneath the bottom of the cone (as a sit-and-wait predator) and waits for insects (preferably ant) to be trapped in the pit.

The edge of the cone is sharp enough for insects to fall to the bottom of the trap easily. Once the antlion realizes that a prey is in the trap, it tries to catch it. However, insects usually are not caught immediately and try to escape from the trap Judith (2010). In this case, antlions intelligently throw sands towards to edge of the pit to slide the prey into the bottom of the pit. When a prey is caught into the jaw, it is pulled under the soil and consumed. After consuming the prey, antlions throw the leftovers outside the pit and amend the pit for the next hunt.

Another interesting behavior that has been observed in life style of antlions is the relevancy with respect to size of the trap and two things: level of hunger and shape of the moon. Antlions tend to dig out larger traps as they become more hungry Judith (2010) and/or when the moon is full. They have been evolved and adapted this way to improve their chance of survival. It also has been discovered that an antlion does not directly observe the shape of the moon to decide about the size of the trap, but it has an internal lunar clock to make such decisions Pisula (2009).

The main inspiration of the ALO algorithm comes from the foraging behaviour of antlion's larvae. Antlions belongs to Myrmeleontidae family and live in two phases of larvae and adult. During their larvae phase, antlions make a small cone shaped trap in order to trap ants. Antlions sit under the pit and wait for ants to be trapped. After feeding on trapped ants, antlions throw the leftovers outside the pit and amend the pit for the next hunt. It has been observed that antlions dig a bigger bit when they are hungry, and this is the main concept in ALO optimization algorithm Mirjalili (2015a).

During optimization, the following conditions are applied:

- Ants move around the search space using different random walks.
- Random walks are applied to all the dimension of ants.
- Random walks are affected by the traps of antlions.
- Antlions can build pits proportional to their fitness (the higher fitness, the larger pit).
- Antlions with larger pits have the higher probability to catch ants.
- Each ant can be caught by an antlion in each iteration and the elite (fittest antlion).
- The range of random walk is decreased adaptively to simulate sliding ants towards antlions.
- If an ant becomes fitter than an antlion, this means that it is caught and pulled under the sand by the antlion.
- An antlion repositions itself to the latest caught prey and builds a pit to improve its change of catching another prey after each hunt.

3.3   Model Framework

In the proposed MoFAL framework as detailed in Algorithm 1, it is assumed that the candidate solutions are moths and the problem's variables are positions of moths in the space dimensions. Therefore, the moths can fly in 1-D, 2-D, 3-D, or hyper dimensional space with changing their position vectors. Since the MFO algorithm is a population-based algorithm, the set of moths is represented in a matrix as follows:

$$\mathbf{M} = \begin{bmatrix} M_{1,1} & M_{1,2} & M_{1,3} & \ldots & M_{1,d} \\ M_{2,1} & M_{2,2} & M_{2,3} & \ldots & M_{2,d} \\ \cdots\cdots\cdots\cdots\cdots\cdots\cdots \\ M_{n,1} & M_{n,2} & M_{n,3} & \ldots & M_{n,d} \end{bmatrix} \tag{1}$$

where n is the number of moths and d is the number of variables (dimension). For all the moths, we also assume that an array exists which holds the related fitness solutions of the moths and it is given by the following equation:

$$\mathbf{OM} = \begin{bmatrix} OM_1 \\ OM_2 \\ \vdots \\ \vdots \\ OM_n \end{bmatrix} \tag{2}$$

Where n denotes the number of moths. Another important characteristics of the algorithm are the moth-flames and is given by the below equation:

$$\mathbf{F} = \begin{bmatrix} F_{1,1} & F_{1,2} & F_{1,3} & \ldots & F_{1,d} \\ F_{2,1} & F_{2,2} & F_{2,3} & \ldots & F_{2,d} \\ \cdots\cdots\cdots\cdots\cdots\cdots\cdots \\ F_{n,1} & F_{n,2} & F_{n,3} & \ldots & F_{n,d} \end{bmatrix} \tag{3}$$

For the flames, the corresponding fitness values are stored in the below matrix:

$$\mathbf{OF} = \begin{bmatrix} OF_1 \\ OF_2 \\ \vdots \\ \vdots \\ OF_n \end{bmatrix} \tag{4}$$

Similarly for ALO, the algorithm mimics the interaction between the antlions and ants in the trap. In its mathematical model, ants are required to move over the search space, and antlions are allowed to hunt them and become fitter using traps. Since ants move stochastically in nature when searching for food, a random walk is chosen for modelling ants movement and it is given by:

$$X(t) = [0, \Omega(2r(t_1), \Omega(2r(t_2) - 1), \cdots, \Omega(2r(t_n) - 1)] \tag{5}$$

where Ω calculates the cumulative sum, n is the maximum number of iterations, t shows the step of random walk (iteration, and $r(t)$ is a stochastic function defined by:

$$r = \begin{cases} 1, & \text{if rand} > 0.5 \\ 0, & \text{otherwise} \end{cases}$$

The position of ants are stored and used during optimization in the matrix equation:

$$\mathbf{M_{ant}} = \begin{bmatrix} A_{1,1} & A_{1,2} & A_{1,3} & \dots & A_{1,d} \\ A_{2,1} & A_{2,2} & A_{2,3} & \dots & A_{2,d} \\ \dots & \dots & \dots & \dots & \dots \\ A_{n,1} & A_{n,2} & A_{n,3} & \dots & A_{n,d} \end{bmatrix} \tag{6}$$

The corresponding fitness function are stored in following equation in order to evaluate the fitness of each ant.

$$\mathbf{M_{OA}} = \begin{bmatrix} f([A_{1,1} \; A_{1,2} \; A_{1,3} \; \dots \; A_{1,d}]) \\ f([A_{2,1} \; A_{2,2} \; A_{2,3} \; \dots \; A_{2,d}]) \\ \dots \\ f([A_{n,1} \; A_{n,2} \; A_{n,3} \; \dots \; A_{n,d}]) \end{bmatrix} \tag{7}$$

We assume that antlions are hiding at some places in the search space and their positions and fitness solutions are saved in the matrix form:

$$\mathbf{M_{antlion}} = \begin{bmatrix} Al_{1,1} & Al_{1,2} & Al_{1,3} & \dots & Al_{1,d} \\ Al_{2,1} & Al_{2,2} & Al_{2,3} & \dots & Al_{2,d} \\ \dots & \dots & \dots & \dots & \dots \\ Al_{n,1} & Al_{n,2} & Al_{n,3} & \dots & Al_{n,d} \end{bmatrix} \tag{8}$$

where $M_{antlion}$ is the matrix for saving the position of each antlion, $Al_{i,j}$ shows the jth dimension's value of ith antlion, n is the number of antlions, and d is the number of variables (dimension).

M_{OAL} is the matrix for saving the fitness of each antlion, $Al_{i,j}$ shows the jth dimension's value of ith antlion, n is the number of antlions, and f is the fitness function.

$$\mathbf{M_{OAL}} = \begin{bmatrix} f([Al_{1,1} \; Al_{1,2} \; Al_{1,3} \; \dots \; Al_{1,d}]) \\ f([Al_{2,1} \; Al_{2,2} \; Al_{2,3} \; \dots \; Al_{2,d}]) \\ \dots \\ f([Al_{n,1} \; Al_{n,2} \; Al_{n,3} \; \dots \; Al_{n,d}]) \end{bmatrix} \tag{9}$$

In MoFAL, we extract the elitism characteristics of ALO algorithm and harmonize it into MFO algorithm. Elitism is an important characteristic of evolutionary algorithms that allows them to maintain the best fitness(s) obtained at any stage of optimization process. In ALO, the best antlion obtained so far in each iteration is saved and considered as an elite. Since elite is the fittest antlion, it can influence the movements of all ants during iteration. Therefore, we assume that every ants random walks around a selected antlion by the roulette wheel and the elite simultaneously as given by the equation:

$$Ant_i^t = \frac{R_A^t + R_E^t}{2} \tag{10}$$

Data: W_n–Number of search agents(N) = 50, Iter$_{MAX}$ – total number of iterations(500) more than 0, Number of variables (d) = 30, $t \in [-1, 1]$, r linearly decrease from -1 to -2

Result: Optimal MFO-ALO flame value within given Design Space.

begin

    Initialize set of moths matrix - Eq. 1

    Calculate corresponding moth fitness value - Eq 2

    Initialize position of ants - Eq. 6

    Evaluate fitness of each ants - Eq. 7

    Find best fitness antlion and assume it as Elite.

    Set **t**:=0

    repeat

        Update flame _no using Eq. 19

        if $t == 1$ **then**

            F = sort(M). Eq 3

            OF = sort(OM), Eq 4

        end

        else

            $F = sort(M_{t-1}, M_t)$

            $OF = sort(M_{t-1}, M_t)$

        end

        forall the *moths/ants* $\in$ *N* **do**

            forall the *parameters/variables* $\in$ *d* **do**

                if *moths/ants* $\leq$ *Flame_no* **then**

                    Tag the best moth position by flames.

                    Update *flame number*, t and r parameters.

                    Calculate moth D using Eq 20

                    Update the matrix M using Eq15 and Eq18

                end

                if *moths/ants* $\geq$ *Flame_no* **then**

                    Tag the best moth position by flames.

                    Update *flame number*, t and r parameters.

                    Calculate moth D using Eq 20

                    Update the matrix M using Eq15 and Eq18

                    Select an antlion using roulette wheel.

                    Update c and d using equation 16 and Eq. 17

                    Create random walk and normalize it. 5

                    Update elite antlion and compare with moth fitness with Eq10

                    Update best antlion/moth position.

                end

            end

            Update elite if an antlion becomes fitter than the elite

        end

        Calculate all fitness values

        Update flames

    until $t \leq Iter_{MAX}$

    Return **Best Flame/antlion** position

end

Algorithm 1: MoFAL Framework

where R_A^t is the random walk around the antlion selected by the roulette wheel at t^{th} iteration, R_E^t is the random walk around the elite at t^{th} iteration, and Ant_i^t indicates the position of ith ant at t^{th} iteration.

The ALO algorithm can be deduced to a three-tuple function that search for global minimum for optimization as follows:

$$ALO(A, B, C) \tag{11}$$

where A is a function that generates the random initial solutions, B manipulates the initial population provided by the function A, and C returns true when the end criterion is satisfied. The functions A, B, and C are defined as follows:

$$\theta \xrightarrow{\text{A}} \{M_{Ant}, M_{OA}, M_{Antlion}, M_{OAL}\} \tag{12}$$

$$\{M_{Ant}, M_{Antlion}\} \xrightarrow{\text{B}} \{M_{Ant}, M_{Antlion}\} \tag{13}$$

$$\{M_{Ant}, M_{Antlion}\} \xrightarrow{\text{C}} \{true, false\} \tag{14}$$

where M_{Ant} is the matrix of the position of ants, $M_{Antlion}$ includes the position of antlions, M_{OA} contains the corresponding fitness of ants, and M_{OAL} has the fitness of antlions.

It is interesting to note that, MFO is also three-tuple function that approximates the global optimal of optimization problems. This uncanny similarity between ALO and MFO serves as one of the basis of our hybridization. Furthermore, the position of each moth is updated with respect to a flame using the following equation:

$$M_i = S(M_i, F_j) \tag{15}$$

where M_i indicate the ith moth, F_j indicates the jth flame, and S is the spiral function.

For MoFAL, in addition to navigation method of moths, we also create the random walk of antlions and normalize it using following equations:

$$c^t = \frac{c^t}{I} \tag{16}$$

$$d^t = \frac{d^t}{I} \tag{17}$$

where I is a ratio, c^t is the minimum of all variables at t^{th} iteration, and d^t indicates the vector including the maximum of all variables at t^{th} iteration.

Elite antlion position is calculated and is compared with that of moth position as given by the logarithmic spiral equation below:

$$\mathbf{X}(t + 1) = \mathbf{D}'. \, e^{bl}. \cos(2\pi l) + \mathbf{X} * (t) \tag{18}$$

where $\mathbf{D}' = [D_i]$ indicates the distance of the ith moth for the jth flame, b is a constant for defining the shape of the logarithmic spiral, and t is a random number in $[-1, 1]$. Where r is linearly decreased from -1 to -2 over the course of iteration.

Flame number is calculated by following equation:

$$Flame\_no = round\{(N - Iteration)\frac{N-1}{MaxIteration}\} \qquad (19)$$

where N is the number of search agent. D is calculated as follows:

$$D_i = |F_j - M_i| \qquad (20)$$

If Elite antlion fitness is greater than that of moth-flame fitness; position vectors are updated using Eq. 10 using both moth flame position and elite antlion position.

4   Simulation Results

4.1   Initialization

A fitness function is a particular type of objective function that is used to summarise, as a single figure of merit, how close a given design solution is to achieving the set aims. Fitness functions are used in evolutionary algorithms to guide simulations towards optimal design solutions. The test functions utilized for bench-marking are basically minimization functions and can be divided into four groups: unimodal, multimodal, fixed dimension multimodal and composite functions.

Unimodal functions defined in Table 1 have only one global optimum. These functions assist in evaluating the exploitation capability of experimented meta-heuristic algorithms.

Table 1. Unimodal Fitness Functions

Function	Dim	Range	f_{min}
$f_1(x) = \sum_{i=1}^{n} x_i^2$	30	$[-100,100]$	0
$f_2(x) = \sum_{i=1}^{n} |x_i| + \Pi_{i=1}^{n}|x_i|$	30	$[-10,10]$	0
$f_3(x) = \sum_{i=1}^{n}(\sum_{j-1}^{i} x_j^2)$	30	$[-100,100]$	0
$f_4(x) = max_i(|x_i|, 1 \le i \le n)$	30	$[-100,100]$	0
$f_5(x) = \sum_{i=1}^{n-1}[100(x_{i+1}-x_i^2)^2 + (x_i - 1)^2]$	30	$[-30,30]$	0
$f_6(x) = \sum_{i=1}^{n}[((x_i + 0.5)^2]$	30	$[-100,100]$	0
$f_7(x) = \sum_{i=1}^{n} ix_i^4 + random[0, 1]]$	30	$[-1.28,1.28]$	0

The fixed-dimension multimodal functions defined in Table 2 include many local optima whose number increases exponentially with the problem size (number of design variables). Therefore, these kind of test problems prove to be useful if the purpose is to evaluate the exploration capability of an optimization algorithm. Therefore this makes the multimodal functions more suitable for benchmarking the exploration ability of an algorithm. Unlike unimodal functions

Table 2. Fixed-dimension Multimodal Fitness Functions

Function	Dim	Range	f_{min}
$f_{14}(x) = (\frac{1}{500} \sum_{j=1}^{25} \frac{1}{j + \sum_{i=1}^{2} (x_i - a_{ij})^6})^{-1}$	2	$[-65,65]$	1
$f_{15}(x) = \sum_{i=1}^{11} (a_i - \frac{x_1(b_i^2 + b_i x_2)}{b_i^2 + b_i x_3 + x_4})^2$	4	$[-5,5]$	0.00030
$f_{16}(x) = 4x_1^2 + 2.1x_1^4 + \frac{1}{3}x_1^6 - 4x_2^2 + x_1 x_2 + 4x_2^4)$	2	$[-5,5]$	-1.0316
$f_{17}(x) = \sum_{i=1}^{4} c_i \exp(-\sum_{j=1}^{3} a_{ij}(x_j - p_{ij})^2)$	2	$[1,3]$	-3.86
$f_{18}(x) = \sum_{i=1}^{5} [t(X - a_i)(X - a_{it})^T + c_i]^{-1}$	4	$[0,10]$	-10.1532
$f_{19}(x) = \sum_{i=1}^{10} [t(X - a_i)(X - a_{it})^T + c_i]^{-1}$	4	$[0,10]$	-10.5363

(Table 1), it can be clearly observed that multimodal functions defined in Table 3 have many local optima with the number increasing exponentially with dimension.

The last set of fitness functions in Table 4 is a combination of various constrained benchmark functions which are shifted, expanded, rotated and integrated in order to provide greater complexity in terms of testing the exploitation, exploration and effectiveness of optimization algorithms.

4.2 Comparative Analysis

A graphical analysis on hybrid algorithms with respect to their parent algorithms are examined in this section. The hybrid algorithm along with their parent algorithms are run simultaneously on 23 constrained bench-marking functions namely - unimodal, multimodal, fixed dimension multimodal and composite functions in order to graphically analyze the rate of convergence of specified algorithms and visually validate their performance with respect to their exploitation and exploration capability. All these algorithms are run for 500 iterations and rate of convergence of all the algorithms are observed.

Table 3. Multimodal Fitness Functions f_8, f_9, f_{10}, f_{11} and f_{13}

Function	Dim	Range	f_{min}
$f_8(x) = \sum_{i=1}^{n} -x_i \sin(\sqrt{|x_i|})$	30	$[-500,500]$	-418.9289
$f_9(x) = \sum_{i=1}^{n} [x_i^2 - 10\cos(2\pi x_i) + 10$	30	$[-5.12,5.12]$	0
$f_{10}(x) = -20 \exp -0.2\sqrt{\frac{1}{n}\sum_{i=1}^{n} x_i^2} - \exp(1n \sum_{i=1}^{n} \cos(2\pi x_i)) + 20 + e$	30	$[-32,32]$	0
$f_{11}(x) = \frac{1}{4000} \sum_{i=1}^{n} x_i^2 - \Pi_{i=1}^{n} \cos(\frac{x_i}{\sqrt{i}}) + 1)$	30	$[-600,600]$	0
$f_{12}(x) = 0.1[\sin^2(3\pi x_i) + \sum_{i=1}^{n}(x_i - 1)^2 + [1 + \sin(3\pi x_i + 1] + (x_n - 1)^2 + [1 + \sin^2(2\pi x_n)]] + \sum_{i=1}^{n} u(x_i, 5, 100, 4)$	30	$[-50,50]$	0
$f_{13}(x) = \sum_{i=1}^{n} \sin(x_i).(\sin(\frac{i.x_i^2}{\pi}))^{2m}, m = 10$	30	$[0,\Pi]$	-4.687

Figure 1 shows the line graphs of convergence of MoFAL hybrid algorithm with respect to their parent MFO and ALO algorithms. From the graph, it can be

Table 4. Composite Fitness Functions

Function	Dim	Range	f_{min}
f_{20}(CF1): $f_1, f_2, f_3, f_4, \ldots, f_{10} = Sphere function$ $\beta_1, \beta_2, \beta_3, \ldots, \beta_{10}] = [1, 1, 1, \cdots, 1]$ $[\lambda_1, \lambda_2, \lambda_3, \ldots, \lambda_{10} = [5/100, 5/100, 5/100, \cdots, 5/100]$	10	$[-5,5]$	0
f_{21}(CF2): $f_1, f_2, f_3, f_4, \ldots, f_{10} = Griewank's function$ $\beta_1, \beta_2, \beta_3, \ldots, \beta_{10}] = [1, 1, 1, \cdots, 1]$ $[\lambda_1, \lambda_2, \lambda_3, \ldots, \lambda_{10} = [5/100, 5/100, 5/100, \cdots, 5/100]$	10	$[-5,5]$	0
f_{22}(CF3): $f_1, f_2 = Ackley's function$ $f_3, f_4, = Rastrigin's function$ $f_5, f_6, = Weierstra's function$ $f_7, f_8, = Griewank's function$ $f_9, f_{10}, = Sphere function$ $\beta_1, \beta_2, \beta_3, \ldots, \beta_{10}] = [1, 1, 1, \cdots, 1]$ $[\lambda_1, \lambda_2, \lambda_3, \ldots, \lambda_{10} =$ $[[5/32, 5/32, 1, 1, 5/0.5, 5/0.5, 5/100, 5/100, 5/100, 5/100]$	10	$[-5,5]$	0
$f_{23}(CF4)$: $f_1, f_2 = Ackley's function$ $f_3, f_4 = Rastrigin's function$ $f_5, f_6 = Weierstra's function$ $f_7, f_8 = Griewank's function$ $f_9, f_{10} = Sphere function$ $\beta_1, \beta_2, \beta_3, \ldots, \beta_{10}] = [1, 1, 1, \cdots, 1]$ $[\lambda_1, \lambda_2, \lambda_3, \ldots, \lambda_{10} = [0.1 * 1/5, 0.2 * 1/5, 0.3 * 5/0.5, 0.4 *$ $5/0.5, 0.5 * 5/100,$ $0.6 * 5/100, 0.7 * 5/32, 0.8 * 5/32, 0.9 * 5/100, 1 * 5/100]$	10	$[-5,5]$	0

observed that MoFAL has the best performance for this multimodal benchmark function F_9. MoFAL achieves a final fitness value of 72.638 followed by ALO at 103.492, whereas, MFO was able to obtain a value of only 201.976. We utilize multimodal functions to evaluate the exploration capability of the investigated meta-heuristic algorithms. Exploration consists of probing a much larger portion of the search space with the hope of finding other promising solutions that are yet to be refined. This operation leads to diversifying the search in order to avoid getting trapped in a local optimum. The graphical analysis clearly demonstrates immaculate exploration capability of MoFAL to jump outside comparatively faster from a local optimum.

4.3 Statistical Analysis

Statistical testing is a process of making quantitative decisions about a problem in which statistical data set is evaluated and taken which is then compared hypothetically Wilcoxon et al. (1963) Wilcoxon, Katti, and Wilcox. In this anal-

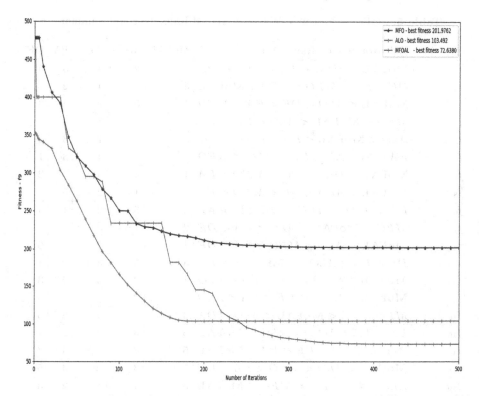

Fig. 1. Comparison of MFO, ALO and MoFAL optimization algorithms for the f9 constrained multimodal benchmark function in 500 iterations.

ysis we consider that there are instances in which is very essential to carry out analysis along the search.

This way, optimization algorithms can be evaluated depending on its convergence performance, which would help when deciding which algorithms perform better among a set of methods that are assumed as equal when only the results at the end of the search are considered. In addition to hybrid algorithms and their parent algorithms, two evolutionary algorithms such as Flower Pollination algorithm (FA) Yang (2012) and Differential Evolution algorithm (DE) Storn and Price (1997) is also considered in order to validate the performance of hybrid algorithm realized.

4.3.1 Wilcoxon Signed Rank Test

The Wilcoxon signed rank test is a non parametric test which means the population data is non-normally distributed. It is solely based on the order of the sample's observations Wilcoxon et al. (1963) Wilcoxon, Katti, and Wilcox. The one with lowest rank will be considered as the best amongst all and vice-versa.

The results of Wilcoxon signed rank test for MoFAL is given in Table 5. The results shows that MoFAL have the lowest rank among all five algorithms

Table 5. MoFAL - Pair-wise Wilcoxon Signed Rank Test and Summary

Function	Wilcoxon Signed Rank Order	MoFAL	MFO	ALO	FA	DE
f_1	**MoFAL** $< MFO < FA < ALO < DE$	1	2	4	3	5
f_2	$DE < FA < MFO < ALO <$ **MoFAL**	5	3	4	2	1
f_3	**MoFAL** $< ALO < DE < FA < MFO$	1	5	2	4	3
f_4	$MFO <$ **MoFAL** $< ALO < DE < FA$	2	1	3	5	4
f_5	$ALO <$ **MoFAL** $< FA < DE < MFO$	2	5	1	3	4
f_6	$FA <$ **MoFAL** $< DE < ALO < MFO$	2	5	4	3	1
f_7	**MoFAL** $< DE < ALO < MFO < FA$	1	4	3	5	2
f_8	**MoFAL** $< ALO < DE < MFO < FA$	4	1	2	5	3
f_9	$FA < ALO < MFO < DE <$ **MoFAL**	5	3	2	1	4
f_{10}	$MFO <$ **MoFAL** $< ALO < FA < DE$	2	1	3	4	5
f_{11}	**MoFAL** $< ALO < MFO < FA < DE$	1	3	2	4	5
f_{12}	$DE < FA < ALO <$ **MoFAL** $< MFO$	4	5	3	2	1
f_{13}	**MoFAL** $< MFO < DE < FA < ALO$	1	2	5	4	3
f_{14}	**MoFAL** $< MFO < DE < FA < ALO$	1	2	5	4	3
f_{15}	$MFO < FA <$ **MoFAL** $< ALO < DE$	3	1	4	2	5
f_{16}	$DE < FA < MFO < ALO <$ **MoFAL**	5	3	4	2	1
f_{17}	$ALO < MFO < DE < FA <$ **MoFAL**	5	2	1	4	3
f_{18}	**MoFAL** $< ALO < MFO < FA < DE$	1	3	2	4	5
f_{19}	$ALO < FA < DE < MFO <$ **MoFAL**	5	4	1	2	3
f_{20}	**MoFAL** $< ALO < DE < FA < MFO$	1	5	2	4	3
f_{21}	$FA <$ **MoFAL** $< MFO < DE < ALO$	2	3	5	1	4
f_{22}	$FA <$ **MoFAL** $< ALO < DE < MFO$	2	5	3	1	4
f_{23}	**MoFAL** $< ALO < DE < MFO < FA$	1	4	2	5	3
Total		57	72	67	74	75

subjected to the statistical testing. Their parent algorithms ALO and MFO competed with MoFAL closely and was ranked second and third respectively. FA and DE algorithms were placed fourth and fifth in order.

The superior performance of MoFAL does not imply it is better than the other algorithms selected for the case study which will lead to the violation of 'free lunch theorem'Ho and Pepyne (2002). Its performance outlines its superiority in selected benchmark functions comparatively.

4.3.2 Kolmogorov-Smirnov Test

The Kolmogorov–Smirnov (K–S test) performs a non-parametric test of the equality of continuous, one-dimensional probability distributions that can be used to compare a sample with a reference probability distribution. In KS test,

Table 6. MoFAL - Kolmogorov-Smirnov test p-value results

Function	MoFAL	MFO	ALO	FA	DE
f_1	N/A	0.0112	**0.0511**	0.0018	0.0018
f_2	N/A	N/A	0.0081	**0.217**	0.0018
f_3	**0.0522**	0.0031	0.0029	0.0012	0.0018
f_4	**.0611**	N/A	N/A	0.0018	**0.0871**
f_5	**.0508**	0.0041	**0.0766**	0.00183	0.00182
f_6	N/A	**0.0691**	N/A	0.8122	2.971
f_7	0.0456	**0.0502**	0.0355	0.0022	**0.0508**
f_8	**0.0722**	N/A	N/A	**0.5001**	N/A
f_9	N/A	0.0022	0.0081	0.0047	N/A
f_{10}	**0.0612**	0.0133	**0.0532**	N/A	0.0017
f_{11}	**0.0711**	**0.0812**	N/A	0.0041	0.0016
f_{12}	0.0341	0.0022	**0.0073**	0.0013	0.0017
f_{13}	0.0302	0.0181	N/A	N/A	N/A
f_{14}	0.0416	N/A	0.0111	0.0018	0.0021
f_{15}	N/A	**0.1629**	0.0011	0.0018	0.0018
f_{16}	**0.0645**	0.0022	**0.0559**	0.0047	N/A
f_{17}	**0.0645**	0.0081	0.0083	0.0047	N/A
f_{18}	N/A	**0.5316**	0.0044	**0.0551**	0.0011
f_{19}	**0.0578**	0.0481	0.0012	**0.0922**	N/A
f_{20}	N/A	N/A	0.0022	0.0012	0.0018
f_{21}	**0.0589**	**0.0719**	0.0018	N/A	**0.0699**
f_{22}	**0.0806**	N/A	0.0299	0.0032	**0.0050**
f_{23}	**0.1055**	N/A	**0.5611**	0.0031	0.0018

we have experimental dataset and generated dataset and the KS test tries to determine whether if two datasets differ significantly. The KS-test has the advantage of making no assumption about the distribution of data. The output of KS test gives the distance D and the $p-$value. A low $p-$value means that it is likely that the data is different from the model. We require a high $p-$value in order to have a good fitting.

The p values of Kolmogorov-Smirnov test over 10 runs are given in Table 6 for MoFAL. Any p values >0.05 confirm the statistical significant superiority of the results. The p values of Table 6 suggest that the superiority of the MoFAL algorithm is statistically significant in the majority of times. In addition, p values show that the MoFAL algorithm rejects the null hypothesis on F16 and F17, validating the competitive results of this algorithm.

4.3.3   Mean-Average Method

This is the simplest and most popular method used for statistical testing of optimization algorithm. In order to have an accurate evaluation of hybrid algorithm, all algorithms selected for statistical testing is run 30 times. Then we take the mean average of each optimization algorithm in order to validate the results. Comparison results for MoFAL are given in Table 7.

According to Table 7, MoFAL were able to find better solutions in 14 out of 23 benchmark functions than the other selected optimization algorithms. ALO provided best global optimum in 8 benchmark functions followed by MFO and DE in 5 fitness functions. FA algorithm gave fair results but they were significantly low results compared to other algorithms.

Table 7. MoFAL - Comparison results of all constrained benchmark functions

Function	MoFAL	MFO	ALO	FA	DE
f_1	0.4283	0.7812	**0.3001**	0.6718	0.8199
f_2	**0.6912**	0.8192	0.7617	3.216	0.8129
f_3	66.812	71.283	**52.822**	131.211	111.877
f_4	**1.312**	5.461	2.912	2.7145	10.381
f_5	**33.456**	33.901	56.9451	411.21	76.81
f_6	.9812	**0.7712**	1.3391	0.8122	2.971
f_7	1.3328	1.6821	1.8802	**0.9318**	1.3128
f_8	**−8236.21**	−4186.85	−3819.71	−5121.99	−1891.12
f_9	56.288	58.001	61.895	133.655	**50.129**
f_{10}	**1.5127**	5.9216	2.015	7.891	9.712
f_{11}	**0.308**	0.5822	1.0877	1.311	1.978
f_{12}	1.8122	0.9811	1.0366	**0.4612**	2.711
f_{13}	**0.0891**	0.0618	**0.0899**	1.899	0.659
f_{14}	1.011	**0.9980**	**0.9980**	**0.9980**	**0.9980**
f_{15}	**0.1618**	1.2318	3.4112	1.118	0.8652
f_{16}	**−1.0365**	−1.0341	−1.0322	−1.0339	−1.6211
f_{17}	**0.2419**	1.562	**0.2421**	0.3978	0.6522
f_{18}	**2.999**	**2.999**	**3.001**	3.122	**2.981**
f_{19}	−2.8125	−1.0082	−1.6125	−2.0923	**−3.8699**
f_{20}	−2.891	−2.7051	**−3.1501**	−2.9912	−1.3697
f_{21}	**−10.0165**	−8.0621	−4.4187	−7.912	−3.6981
f_{22}	**−10.2081**	−8.6955	−8.6125	−8.1281	−1.6987
f_{23}	**−10.4798**	−10.6365	−3.8112	−9.5211	**−10.4821**

5 Application for Classical Engineering Problems

Engineering design is the method that engineers use to identify and solve problems. In constrained engineering design process, engineers must identify solutions that include the most desired features and fewest negative characteristics. They should also specify the cost functions and their limitations of the given scenario, which could include time, cost, and the physical limits of tools and materials. Constrained engineering design optimization problems are usually computationally expensive due to non-linearity and non convexity of these constraint functions. Evolutionary population based algorithms are widely used to solve constrained optimization problems. Many researchers have implemented many heuristic and meta-heuristic optimization algorithms to solve constrained optimization problems in engineering design.

These meta-heuristic optimization algorithms are of great research interest in recent times due to their ability in finding optimal solutions within short time especially when these real world engineering design problems consists of large number of design variables and multiple constraints which makes the solution search-space larger, complicated and non-linear. Penalty function methods are found to be quite popular due to their simplicity and ease of implementation. In this method, search agents are assigned big objective function values if they violate any of the specified constraints. In this section, we try to solve a real world engineering design problem using hybrid algorithm in order to observe the performance and benchmark the performance when compared to its parent and other popular evolutionary algorithms.

5.1 Pressure Vessel Design Problem Using MoFAL

Pressure vessel design problem is constituted by four decision variables: the thickness of the shell (T_s), the thickness of the head (T_h), the inner radius (R), the length of the cylindrical section without considering the head (L). They are represented by (z_1, z_2, z_3, z_4). The objective function of the problem is to minimize the total cost, including the cost of material, forming, and welding. So the general pressure vessel design optimization model can be expressed as:

$$\text{minimize}_{z} \; f(\mathbf{z}) = 0.6224 z_1 z_2 z_3 z_4 + 1.7781 z_2 z_3^2 + 19.84 z_1^2 z_3$$

$$\mathbf{z} = [z_1, z_2, z_3, z_4] = (T_s)(T_h)(R)(L),$$

$$\text{subject to } h_1(\mathbf{z}) = -z_1 + 0.0193 z_3 \leq 0,$$

$$h_2(\mathbf{z}) = -z_2 + 0.00954 z_3 \leq 0 \tag{21}$$

$$h_3(\mathbf{z}) = -\pi z_3^2 - z_4 - \frac{4}{3}\pi z_3^3 + 1296.000 \leq 0$$

$$h_4(\mathbf{z}) = z_4 - 240 \leq 0$$

where

$$0 \leq z_1 \leq 100,$$

$$0 \leq z_2 \leq 100,$$

$$10 \leq z_3 \leq 200,$$

$$10 \leq z_4 \leq 200$$

Table 8 compares the best solutions for pressure vessel design problem obtained by using MoFAL compared with its parent optimizers MFO and ALO as well as FA and DE optimization algorithm. As per the deduced results in Table 8, it is abundantly clear that MoFAL was significantly a better performer compared to the rest of the optimizers.

Table 8. Pressure Vessel design problem results using MoFAL

Parameters	MoFAL	MFO	ALO	FA	DE
$z_1(T_s)$	0.81250	0.81250	1.1250	0.81360	1.12500
$z_2(T_h)$	0.43750	0.43750	0.43750	0.50010	0.43750
$z_3(R)$	41.9104	42.09771	42.0984	42.09110	41.6399
$z_4(L)$	176.6471	176.5110	176.6372	176.6372	176.6502
$f(\mathbf{z})$(Optimum cost)	6060.0777	6059.7341	6063.0256	6061.0115	6088.8755

6   Conclusion

The article proposed a new hybrid evolutionary optimizer namely MoFAL. MoFAL was developed from Hybrid Moth Flame Optimization and Ant Lion Optimizer algorithms. The resulting hybridization process improved the exploration and exploitation ability of optimization algorithms as depicted in the results and analysis sections. The proposed algorithm was successfully analyzed for validation of performance, rate of convergence and local minima avoidance on 23 different classical constrained benchmark functions. These 23 benchmark fitness functions composed of unimodal, multimodal, fixed-dimension multimodal and composite fitness functions in order to evaluate exploration and exploitation ability of optimization algorithm effectively.

In addition, three different statistical testing were carried out to validate the results. The results demonstrated the superiority of the MoFAL algorithm in solving various constrained and unconstrained problems. Also, the obtained results show that MoFAL acquire significantly improved solutions compared with their parent and other recent and popular optimization algorithms. Moreover, MoFAL was applied to solve a real world engineering design problem with large number of variables and constraints and verified the notion that MoFAL was capable in handling various real world conjunctional optimization problems with lower computational efforts.

7   Future Research

For future work, large scale optimization problems can be studied and realized using MoFAL. A self-adaptive method of choosing parameters can be developed in order to further improve the efficiency of MoFAL. In addition, a multi-objective version of MoFAL can also be initiated and developed which has immense scope in diverse real world optimization applications.

Acknowledgments. The work in this article is supported by the Optimization Problems Research and Application Laboratory (OPR-AL), Ryerson University and Natural Sciences and Engineering Research Council of Canada (NSERC). Also, we thank for the open source datasets provided by the original algorithm formulators at Mirjalili (2020a), Mirjalili (2020b) which assisted in our research experiments.

References

Black, P.: "Gray code", from dictionary of algorithms and data structures, Paul E. Black (ed.). NIST, January 2005

Elkhechafi, M., Hachimi, H., Elkettani, Y.: A new hybrid cuckoo search and firefly optimization. Monte Carlo Methods Appl., 2 (2018)

Frank, K.: Effects of artificial night lighting on moths. In: Ecological Consequences of Artificial Night Lighting, pp. 305–344 (2006)

Ganesan, K., Ponnambalam, S.G., Jawahar, N., Janardhanan, M.N.: An effective hybrid cuckoo search and genetic algorithm for constrained engineering design optimization. Eng. Optim. **46**, 1331–1351 (2014)

Gaston, K., Bennie, J., Davies, T., Hopkins, J.: The ecological impacts of nighttime light pollution: a mechanistic appraisal. Biol. Rev. Camb. Philos. Soc. **88**, 912–927 (2013)

Grosan, C., Abraham, A.: Hybrid evolutionary algorithms: methodologies, architectures, and reviews. In: Abraham, A., Grosan, C., Ishibuchi, H. (eds.) Hybrid Evolutionary Algorithms. Studies in Computational Intelligence, vol. 75, pp. 1–17. Springer, Heidelberg (2007). https://doi.org/10.1007/978-3-540-73297-6_1

Ho, Y.C., Pepyne, D.L.: Simple explanation of the no free lunch theorem of optimization. Cybern. Syst. Anal. **38**(2), 292–298 (2002). https://doi.org/10.1023/A:1016355715164

Holland, J.H.: Genetic Algorithms, vol. 266 (1992)

Hybrid Algorithms -Evolutionary: Hybrid algorithm – Wikipedia, the free encyclopedia (2020). https://en.wikipedia.org/wiki/Hybrid_algorithm. Accessed 29 Oct 2020

Jangir, P., Parmar, S., Jangir, N., Kumar, A., Trivedi, I., Bhoye, M.: Optimal power flow using an hybrid particle swarm optimizer with moth flame optimizer. Appl. Soft Comput. (2016)

Judith, E.: Perspectives on Animal Behavior (2010)

Kennedy, J., Eberhart, R.: Particle swarm optimization, pp. 1942–1948 (1995)

Köse, U.: An ant-lion optimizer-trained artificial neural network system for chaotic electroencephalogram (EEG) prediction. Appl. Sci. **8**, 1613 (2018)

Li, Z., Zhou, Y.Q., Zhang, S., Song, J.: Lévy-flight moth-flame algorithm for function optimization and engineering design problems. Math. Probl. Eng. **2016**, 1–22 (2016)

Mageshkumar, C., Karthik, S., Arunachalam, V.P.: Hybrid metaheuristic algorithm for improving the efficiency of data clustering. Cluster Comput. **22**(1), 435–442 (2018). https://doi.org/10.1007/s10586-018-2242-8

Majhi, S., Biswal, S.: Optimal cluster analysis using hybrid k-means and ant lion optimizer. Karbala Int. J. Mod. Sci. **4**, 347–360 (2018)

Mirjalili, S.: The ant lion optimizer. Adv. Eng. Softw. **83**(C), 80–98 (2015a). https://doi.org/10.1016/j.advengsoft.2015.01.010

Mirjalili, S.: Moth-flame optimization algorithm. Know.-Based Syst. **89**(C), 228–249 (2015b). https://doi.org/10.1016/j.knosys.2015.07.006

Mirjalili, S.: Matlab central file exchange. Retrieved ant lion optimizer (alo) datasets, October 2020a. https://www.mathworks.com/matlabcentral/fileexchange/49920-ant-lion-optimizer-alo

Mirjalili, S.: Matlab central file exchange. Retrieved moth-flame optimization (mfo) datasets, October 2020b. https://www.mathworks.com/matlabcentral/fileexchange/52269-moth-flame-optimization-mfo-algorithm

Parvathi, P., Rajendran, R.: A hybrid FCM-ALO based technique for image segmentation, pp. 342–345, October 2016

Pisula, W.: Curiosity and Information Seeking in Animal and Human Behavior, January 2009

Sarbazfard, S., Jafarian, A.: A hybrid algorithm based on firefly algorithm and differential evolution for global optimization. Int. J. Adv. Comput. Sci. Appl. **7**, 95–106 (2016)

Sayed, G.I., Hassanien, A.E.: A hybrid SA-MFO algorithm for function optimization and engineering design problems. Complex Intell. Syst. **4**(3), 195–212 (2018). https://doi.org/10.1007/s40747-018-0066-z

Scharf, I., Ovadia, O.: Factors influencing site abandonment and site selection in a sit-and-wait predator: a review of pit-building antlion larvae. J. Insect Behav. **19**(2), 197–218 (2006). https://doi.org/10.1007/s10905-006-9017-4

Scharf, I., Subach, A., Ovadia, O.: Foraging behaviour and habitat selection in pit-building antlion larvae in constant light or dark conditions. Anim. Behav. **76**, 2049–2057 (2008)

Stojanovic, I.: Application of heuristic and metaheuristic algorithms in solving constrained weber problem with feasible region bounded by arcs, April 2017

Storn, R., Price, K.: Differential evolution - a simple and efficient heuristic for global optimization over continuous spaces. J. Global Optim. **11**(4), 341–359 (1997). https://doi.org/10.1023/A:1008202821328

Ullah, I., Hussain, S.: An efficient energy management in office using bio-inspired energy optimization algorithms. Processes **7** (2019)

Wahid, F., Ghazali, R., Shah, H.: An improved hybrid firefly algorithm for solving optimization problems. In: Ghazali, R., Deris, M.M., Nawi, N.M., Abawajy, J.H. (eds.) SCDM 2018. AISC, vol. 700, pp. 14–23. Springer, Cham (2018). https://doi.org/10.1007/978-3-319-72550-5_2

Wang, F.: Hybrid optimization algorithm of PSO and cuckoo search. In: 2011 2nd International Conference on Artificial Intelligence, Management Science and Electronic Commerce (AIMSEC), pp. 1172–1175, August 2011

Wilcoxon, F., Katti, S., Wilcox, R.A.: Critical values and probability levels for the Wilcoxon rank sum test and the Wilcoxon signed rank test, Pearl River, N.Y. American Cyanamid, January 1963

Xia, X., et al.: A hybrid optimizer based on firefly algorithm and particle swarm optimization algorithm. J. Comput. Sci. **26** (2017)

Xu, H., Liu, X., Su, J.: An improved grey wolf optimizer algorithm integrated with cuckoo search, vol. 1, pp. 490–493 (2017)

Yang, X.-S.: Flower pollination algorithm for global optimization. In: Durand-Lose, J., Jonoska, N. (eds.) UCNC 2012. LNCS, vol. 7445, pp. 240–249. Springer, Heidelberg (2012). https://doi.org/10.1007/978-3-642-32894-7_27

Analysis and Optimization of Low Power Wide Area IoT Network

Shilpi Verma$^{(\boxtimes)}$, Sindhu Hak Gupta, and Richa Sharma

Department of Electronics and Communication Engineering, ASET, Amity University, Noida, Uttar Pradesh, India

Abstract. LoRa (Long Range) is one of the latest Low power Wide Area Network (LPWAN) technology that has increased the number of IoT applications because of its extended battery life, low data rate and large coverage area. In this paper, we have investigated and analyzed the effects of different transmission parameters of LoRa on the estimated battery life of sensors. To extend the battery life of LoRa based sensors, optimal values of non-constrained parameters such as Spreading factor (SF), Coding rate (CR) and Bandwidth (BW) has been analyzed on the basis of Mean Square Error (MSE) Function. The potency of MSE is evaluated by the means of Artificial Neural Network using neural network fitting tool in MATLAB for simulations. In comparison to current prevalent lifetime of LoRa based node i.e. 10years, the optimization insights an increase in the battery life of nodes upto 19 years. Encapsulating these benefits of LoRaWAN, this technology has been proved as the propitious methods in different and wide applications of IoT.

Keywords: LoRa · Artificial Neural Network · Battery life

1 Introduction

The basic objective of IoT is to "connect the unconnected" [1]. IoT characterizes intelligence, complex system architecture, everything-as-a service, size, time and space considerations and hence anticipates duplex interaction among the devices like sensors, vehicles, home appliances, agricultural products, and other multidisciplinary IoT applications [2]. The three basic layers which jointly form topology of IoT are classified as the application layer, network layer, and perception layer. Bluetooth, ZigBee, Z-Wave, 2G/3G/4G/5G, LoRa, SigFox, Weightless, WAVIoT, and Wi-Fi HaLow - all are the technologies used in the network layer which interfaces the hardware and sensors of perception layer with the different applications in application layer like smart city, precision agriculture in suburban areas etc. [3, 4].

Many IoT infrastructure constitutes resource constrained (RC) devices. They are smaller in size, with less data handling capacity and power limitations. The battery lifetime of nodes in the networks are susceptible to premature failure because of more power consumption than other nodes due to workload variations, non-uniform set-up of communication and heterogeneous hardware requirements. Due to distributed-infrastructure, the nodes operation varies from one location to another and rapid multiplication of sensor

© Springer-Verlag GmbH Germany, part of Springer Nature 2021
M. L. Gavrilova and C. J. K. Tan (Eds.): Trans. on Comput. Sci. XXXVIII, LNCS 12620, pp. 98–112, 2021.
https://doi.org/10.1007/978-3-662-63170-6_6

devices in modern applications results in increasing power consumption [5]. To control the energy, a number of technologies have been developed on the basis of coverage, lifetime and cost [6]. The technologies residing under this canopy are Ultra Narrow band by SigFox, NB-IoT, LoRa, RPMA by Ingenu, 3GPP: LTE-M, LTE-Cat 0 and Weightless-W, N and P. Classification of LPWAN technologies are based on operating frequencies as licensed and unlicensed. The main defiance faced by licensed bands service providers is the possibility of non-mobile service providers to in-take market share by offering IoT infrastructure and based devices without buying expensive spectrum. For resource constrained devices, unlicensed LPWAN technologies which have come into existence are LoRa, SigFox, Ingenu and Weightless. To transmit over long distance, LoRa is suitable as it works in the ISM band at 433, 868 or 915MHz with less energy consumption and low data rate. LoRaWAN protocol targets a trade between low power utilization and increased distance [7].

The present work is focused on LoRa, a modulation technique implemented on Chip Spread Spectrum (CSS) including physical layer specification. The outlined parameters of the LoRa radio which control the overall network performance are categorised as: Spreading Factor (SF), Bandwidth (BW), Coding Rate (CR) and the Carrier frequency. LoRa is capable to control different data rates via the coding rate, spread factor and bandwidth. And hence this helps in reducing the energy consumption, interference and error rates. However, due to increase in the utilization of LoRa in IoT applications, the availability of optimizing its performance parameters are still sparse. The optimized parameters helps in extending the lifetime of the LoRa based nodes. The optimization is performed using nftool which is based on Artificial Neural Network (ANN). The ambition of ANN is to define the Mean Square Error Function which evaluates the minimum error optimized value of the weights and performs the inter-linking among the neurons in different layer of system. The advantages of the ANN are- the ability to learn feed-forward mechanism, flexibility, fast implementation and low estimated error. The three important non-constrained parameters on which the optimization of battery life depends are: Spreading factor (SF), Bandwidth (BW) and Coding rate (CR). Lifetime denotes the number of days/year the node can perform operation without failure and delay [8]. These three prominent factors are the priority for optimization because most of the energy is consumed in data Communication. The main contributions of the paper are:

- Performance evaluation of a LoRa node by applying mathematical model comprising of diverse specifications of the LoRa network like SF, CR and BW.And also encases the effect of parameters on the battery life.
- Optimization of battery life of a LoRa node by considering all the combination of non-constrained parameters.
- Critical comparison of optimized and non-optimized value in respect of time on air, time cycle and energy cycle.

The rest of the work in the paper is categorised under the following headings: Related work in Sect. 2, followed by the study of LoRa and its parameters in Sect. 3. Section 4 includes the system model and the related equations. Section 5 underlines the optimization and its working principle. The outcomes are compiled in Sect. 6 and the conclusion is mentioned in Sect. 7.

2 Related Work

Several recent works has been focused on LoRa and its effect on the transmission configurations over IoT network. The parameters that have been the focal point in recent research are throughput, delay, coverage, energy, path loss and network density. The earlier proposed work derives the intricate SF, bandwidth and transmission power to increase the throughput by manyfold. Many researches have been undertaken regarding the reduction of energy consumption of different classes of nodes by controlling the transmission power [9]. Table 1 summarizes some of the previous work based on performance and computational analysis of nodes, outlined parameters and resource management.

Table 1. Comparison of related works

References	Performance metrics	Highlighted features	Results accomplished
[10]	Power and bandwidth	Proposed a mathematical formula in terms of power and band usage	Throughput increases by 238.8%
[11]	Energy	Accounts the modeling of the Class A sensor node units	Minimization of energy consumption during acknowledgement transmission
[12]	Throughput	Optimizes a global network configuration to maximize throughput	Increase in throughput by 147% in the network of 200 IoT nodes
[13]	Energy and packet loss	Highlights the trade-off between SF and transmission power	Reduction in energy consumption from 30J to 15J by bandit algorithm
[14]	Network density	Behavioral analysis of LoRa networks using message replication	In low density networks replication is useful
[15]	Path Loss	Utilized LoRa to maximize coverage in dense urban area	Gateways and channels are analyzed for multipoint-to-point communications
[16]	Throughput	Increased time-on-air is compensated by the reducing the interference	Throughput increases by 30%
[17]	Energy	Optimizes SF to increase the PRP for the average energy consumption	Current consumption reduces to 63mAh with increase in PRP performance
In proposed work	Battery life	Analyzing the effect of SF, CR, TOA and BW on Battery life and optimizing these parameters to achieve increased battery life	The battery of the node can last upto approx. **19 years** as compared to presently defined lifetime i.e. 10 years

Although, the above described related works have shown a detailed analysis of LoRa and its parameters, but none of them has made an attempt to optimize all the outlined non-constrained parameters to evaluate the extended lifetime of the nodes. Exceptionally, in the paper [17], energy is minimized by optimizing the SF to increase the packet reception rate but it does not define the lifetime of individual nodes that is dependent on the low duty cycle, SF and TOA. It is necessary to select a good transmission parameter configuration so that the energy consumption and performance network is balanced.

The proposed work is different from the previous studies in a way that it focuses on the optimization of various parameters through mean square error based approach. It is aimed to achieve the pre-eminent set of non-constrained parameters that will prioritize the battery life.

3 LoRa and LoRaWAN

The LoRa Alliance is an association formed in 2015. It standardizes the LPWAN with the different specifications of LoRaWAN. It has developed compliance program and a certification that ensures interoperability. The basis of LoRa is chirp spread spectrum modulation technique and it manages the trade-off between low data rate and increased distance. The parameters affecting the LoRaWAN end devices operation and power consumption are assimilated by the performance of Physical layer and Medium Access Control (MAC) layer. The physical layer is designated to assign the data rate, spreading factor, bandwidth and transmission power. The MAC layer illustrates the LoRaWAN battery life optimization [18]. The performance requirements for MAC layer are throughput, stability and energy efficiency. To optimize end applications, LoRaWAN utilizes various device classes. The device class deals with network downlink latency versus battery lifetime.

The physical parameters of a LoRa device required to tune its performance are Spreading Factor (SF), Coding rate (CR), Transmission power (TP) and Bandwidth (BW). The SF decides the carrier of data on specific number of chirps required. It ranges from 7–12. Higher the value of SF, higher will be the Signal to Noise Ratio (SNR). The other parameter, that is, CR adjusts the ratio of bits transmitted to the actual information bits. The coding rate is defined as 4/(4 + CR). Here the value of CR can be set as 1, 2, 3 or 4. The value of CR = 4/5 indicates that, one bit of correction code is added with every four bits of data. On the other hand, the value of CR = 4/8 refers that four correction bits are added with four bits of data. Although, the higher value of CR provides protection, it also shoots up the time on air (TOA). The TP is the power required for transmission of a specific data packet. It ranges from 2dBm to 20dBm. Lastly, the range of frequencies is specified by bandwidth. LoRa operates at 125 kHz, 250 kHz or 500 kHz. With the increase in BW, data rate increases and simultaneously sensitivity decreases due to short TOA. The computation of TOA of LoRa transmission is done according to the combination of several radio parameters: SF, BW, and CR and payload size [19]. These parameters affects the energy consumption notably and the variations can be clearly identified in terms of time cycle (T_{cycle}) and energy cycle (E_{cycle}).

The various applications incorporating LoRaWAN technology in the present scenario includes Water and gas leakage detection, Smart building security, Inspection of safety, Space optimization and many more. Concentrating on health services, the WBAN is made of several biosensors which is either implanted or deployed on the body to measure medical parameters like arterial pressure, body temperature and EEG Signal [20]. Secondly, WBAN is also prominent solution for military actions such as safeguarding. But the depletion of battery life due to continuous use of the optimal sensors leads to a rapid minimization the lifetime. The features of the applications are utilized in day-to-day life for safe and secure living, making long battery life and energy efficiency as the key aspects to meet the applications requirement. The two key aspects of the nodes and their inter-dependence on the radio parameters are analyzed by the mathematical equations as described in the next section.

4 System Model

The useful combinations of parameters need to be selected for evaluating the important details of energy consumption of the end nodes. The energy consumption of single node depends on its active and sleep state. LoRa promises that the long life of a node is only possible through careful selection of parameter configuration and duty cycle. The ratio of time during which device is in active state to the total time (i.e. active state and sleep state) is known as duty cycle. When the sleep state increases, the duty cycle decreases resulting in the reduction of power consumption. To calculate the lifetime, the radio parameters that are utilized are pointed below-

The equivalent bit rate (R_b) in LoRa [21] is given in terms of

$$R_b = SF * \left(\frac{4}{4 + CR}\right) * \left(\frac{1}{T_s}\right) \tag{1}$$

Here, T_s represents the symbol time. The TOA defines the amount of time each device takes for transmission of total packets and is calculated by the physical layer payload size. In LoRa, the payload length varies between 51 bytes-222 bytes.

$$T_{payload} = \left(n_{payload}\right)(T_s) \tag{2}$$

Where, $T_{payload}$ is the TOA of payload (ms) and n $_{payload}$ is the payload length. Similarly, LoRa physical packet contains preamble of 2 symbols of synchronized word and 2.25 symbols of SFD. The preamble duration ($T_{preamble}$) can be calculated using Eq. (3).

$$T_{preamble} = (N_{preamble} + 4.25) * T_s \tag{3}$$

Here, $T_{preamble}$ denotes the preamble duration in (ms), $N_{preamble}$ represents the length of preamble and T_s is the symbol time in (ms). TOA is the transmission of total message bits i.e. a combination of payload duration and preamble duration [20] and is computed with the formula-

$$T_{packet} = TOA = T_{payload} + T_{preamble} \tag{4}$$

By substituting the value of Eq. (2) and (3) in Eq. (4)

$$TOA = [n_{payload} + N_{preamble} + 4.25] * T_s \tag{5}$$

The modified form of TOA on the basis of bit rate [20] can be defined as-

$$TOA = \frac{\left[n_{payload} + N_{preamble} + 4.25\right] * SF * CR}{R_b} \tag{6}$$

The duration for which the device is in silent state T_{off} [22], is calculated by

$$T_{off} = \frac{TOA}{Duty\ Cycle} - (TOA) \tag{7}$$

Here, the duty cycle defines the ratio of active period to total time period of a node. The relation between the T_{off} and TOA is represented in form of T_{cycle} as

$$T_{cycle} = T_{off} + TOA \tag{8}$$

Since duty cycle is the duration of time a node is transmitting, the cycle duration T_{cycle} indicates a fraction of time for which the node is allowed to transmit. Here TOA represents the transmission duration of a packet. The total time taken to complete one cycle of transmission and reception of message bits is given by the formula-

$$T_{cycle} = 100 * \left(\frac{TOA}{Duty\ Cycle} \right) \tag{9}$$

The parameter T_{cycle} is the time duration of a single transmission cycle. Energy parameters 'Energy consumed' and 'Energy cycle' each indicates the energy contained in a battery and energy expended for every transmission. The lifetime of battery in Eq. (9) is assumed as a perfect battery with no degradation [22].

$$\text{Lifetime} = T_{cycle} * \left(\frac{Energy\ consumed}{Energy\ cycle} \right) \tag{10}$$

5 Optimization Using ANN

Neural network fitting tool (nftool) of MATLAB is used to carry out optimization. It has been discovered in (10) that lifetime of the node depends upon the various parameters like time on air, bandwidth, payload length, spreading factor, duty cycle, coding rate, preamble length, energy consumed and symbol time. These parameters have been calculated with the help of LoRa calculator. Spreading factor, coding rate, bandwidth and duty cycle are termed as free parameters or non-constrained parameters because they are non-linear in nature. The optimization has been performed utilizing the free parameters.

In ANN, the network design is illustrated in Fig. 1. The network design is made up of three layers and the three layers are Input layer, Hidden layers and Output layer. The first layer, that is, the input layer sends the input variables to the hidden layer. Hidden layer lies in the middle of input and the output layer. It contains artificial neurons that take a set of weighted inputs. The activation function produces the output. The output layer is responsible for generating output received from hidden layer. The number of neurons in hidden layer has to be chosen carefully to avoid issues like under fitting, over fitting and increase in training time of the network. A set of input is fed which leads to specific output. The comparison is made between the output and the input target, until the network output matches with the target. The goal of ANN is to calculate the Mean Square Function (MSE). To decrease the error of the increased data sample, the average of distribution is taken. For proposed ANN model, spreading factor, bandwidth, coding rate, duty cycle and energy consumed have been used as input parameters and battery life is used as output parameter.

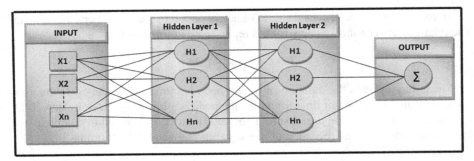

Fig. 1. ANN based "nftool"

A total of 108 samples in the data set have been created to find the optimized value. The hidden layer contains 10 neurons. The created dataset has been bifurcated into training set and testing set. These sets have been further selected for testing and validation purposes. The training set uses 70% of the data and remaining data is used as test set. Neural network is trained to perform the specific function by calibrating the value of weights (connections) between the elements. The performance evaluation of the ANN is represented and analyzed by four different plots- Regression plot, performance plot, training state plot and error histogram graph. The next section portrays the performance analysis of the proposed model with the help of simulation results.

6   Results and Discussions

The effectiveness and the inter-dependence of the LoRa transmission parameters is numerically simulated with the help of LoRa Calculator Tool developed by Semtech. This software is based on LoRa modem and allows quick evaluation of several LoRa configurations by simplifying design decisions (Table 2).

Table 2. Simulation parameters

Parameters	Value
Carrier frequency	865 MHz
Spreading factor	7 to 12
Bandwidth	125 kHz, 250 kHz, 500 kHz
Coding rate	4/8, 4/7 and 4/6
Payload length	8 bytes
Programmable preamble	8 symbols
Total Preamble length	10.25 symbols
Transmission power	17 dBm
Battery capacity	1000 mAh

It is known that LoRa sensors operate for many years without a constant power supply. The energy consumption of different states depends on selected SF and BW. The goal of our objective is to analyze the effect of variables in the considered scenario for extended battery evaluation of the system along with energy efficiency. The lifetime factors of the nodes are TOA, T_{cycle} and E_{cycle}. The various effects of the transmission parameters on the lifetime factors has been analyzed graphically and the observed results are utilized as the base parameters in the optimization.

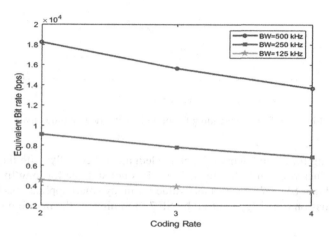

Fig. 2. Effect of coding rate and bit rate considering variable bandwidth

Figure 2 shows the effect of different CR and bandwidth on bit rate. It is analyzed that at a particular BW, when the CR increases then equivalent bit rate decreases due to increase in TOA. Secondly, as the BW increases i.e. more data is received at the same time, the bit rate also escalates. For the present IoT applications, LoRa can be fruitful because as the bit rate increases rapidly with increase in BW, at a particular SF, it results in high speed of data transmission with less number of redundant message bits.

Similarly, Fig. 3 presents the effect of SF and CR on TOA. It is concluded that as the SF increases at a specific CR, the value of TOA also increases because the increase in number of chips in a symbol leads to more time in processing gain at receiver side. This reduces the Battery Life of the devices and increases the energy consumption. It is required to examine that while sending a LoRaWAN message, most of the time is

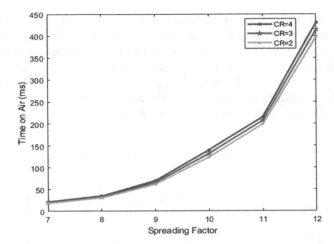

Fig. 3. Effect of Spreading factor and coding rate on time on air

spent on waiting for receive windows acknowledgment. Secondly, with the increase in CR, the TOA increases. In LoRa, the higher SF is not suitable for long life devices. It is generalized that SF = 7 is acceptable value for many smart applications because the lowest transmission time is guaranteed by SF 7. It minimizes the interference.

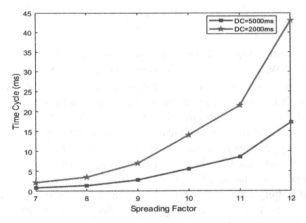

Fig. 4. Effect of spreading factor and duty cycle on time cycle

The term 'time cycle' is directly proportional to the TOA and inversely proportional to the duty cycle. In LoRa Calculator, the duty cycle is calculated in terms of 'ms'. Figure 4 depicts the effect of duty cycle and CR on time cycle. It is noticeable that as the spreading factor increases, time cycle increases hence shooting up the energy consumption. Also, as the duty cycle decreases at a particular SF, the time cycle increases indicating that the device is in OFF state when not required. The energy cycle represents the total energy consumption during transmission and reception of the message.

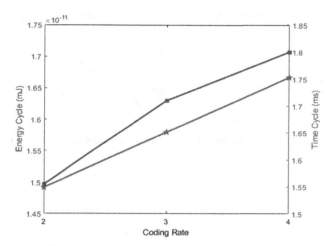

Fig. 5. Effect of coding rate on time cycle and energy cycle

On the other hand, Fig. 5 presents the effect of different CR on time cycle and energy cycle. It is noticed that as the coding rate increases, the energy cycle and time cycle also increases. This means that the increase in number of redundant bits increases the TOA and hence both the cycle also increases.

As discussed earlier, the data set containing 108 samples has been created by taking various combinations of non-constrained parameters of LoRa. The impact of all the parameters is analyzed graphically. It has been discovered that with the minimum SF and higher bandwidth, the lifetime of the node increases. Keeping this in mind, ANN has been implemented for optimization, considering SF as the base parameter associated with the bandwidth. Figure 6 represents the regression curve fit of input and output dataset. Regression plot tracks the targets for training, testing and validation with respect to output. The value of R should be less than 1.To evaluate the optimal extended battery life, an optimized value of free-parameters is obtained using the mean squared error plot. Figure 7 depicts the minimum error that has been observed using the mean square error graph. It highlights the coordinates of the optimized value. The x-coordinate represents the input sample number and corresponding to it the y-coordinate shows the minimum error value.

With the help of coordinates of the minimum error, the optimized value of all the input parameters and the output parameter has been analyzed and listed in Table 3. The IoT applications using LoRa based sensor is battery constraint. It is not feasible to

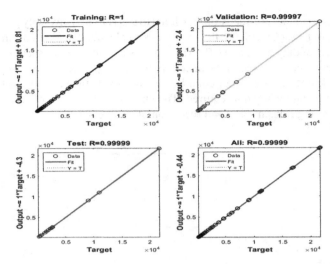

Fig. 6. Regression and error histogram plot of nftool for training and testing of network

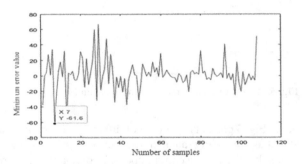

Fig. 7. Mean squared error plot

replace or recharge the battery frequently. Therefore, power and energy optimization is required to make an energy efficient network.

Table 3. Optimized value of transmission parameters

Parameters	Optimized value
Spreading factor (SF)	7
Coding Rate (CR)	4/8
Bandwidth (BW)	250 kHz
Duty Cycle (DC)	2000 ms

It has been analyzed that the maximum battery life to be accommodated with optimized parameters is **7051.23 days (approximately 19 years)**. The resultant optimized value of the parameters of LoRa are- SF = 7, BW = 250 kHz, CR = 4/8 and duty cycle = 2000 ms. All of these diverse parameters play an important role in analyzing the lifetime of the nodes as they determine the time duration of a packet and the amount of energy consumed during the transmission. It is also evident that the energy cycle of the optimized value is less than the average value of non-optimized data samples. Hence, the low energy consumption of the optimized value has turned out to be beneficial and efficient in future for various IoT applications.

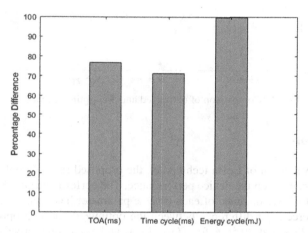

Fig. 8. Percentage difference between optimized and non-optimized value

Figure 8 presents the percentage difference between the optimized and non-optimized values. The percentage difference value of TOA, T_{cycle} and E_{cycle} has been recorded as 76.75%, 71.23 and 99.56% respectively. Correspondingly, in Fig. 9 the comparison of the optimized and non-optimized values has been represented graphically in terms of lifetime factors. The results show that the evaluated optimized value of transmission parameters has proved to be better than conventional values with minimum time on air and minimum energy utilization. This also demonstrates that optimization method proposed on the LoRa transmission parameters is efficient in relation to the other optimization methods presented.

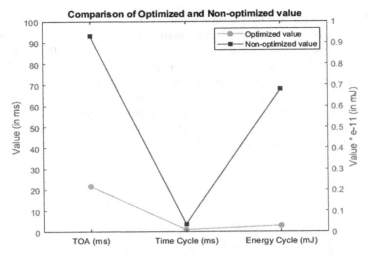

Fig. 9. Comparison of optimized and non-optimized value

7 Conclusion

Based on the evaluation of LoRa technology, the proposed research enlists the effects of LoRaWAN features on the device performance. The performance has been evaluated graphically and the significance of each diverse parameter has been observed. Results highlight the percentage difference between the optimized and non-optimized values as 76.75%, 71.23% and 99.56% for TOA, T_{cycle} and E_{cycle} respectively. In addition, the optimization has been achieved with the help of ANN and it was concluded that the estimated battery life can be extended upto approximately 19 years. The real time IoT applications require extended battery life and minimal power consumption. Since IoT deployment acknowledges "heterogeneity", a large number of devices are needed that can increase the scalability, stability and reliability of the system. Therefore, to implement innovative energy efficient IoT applications, an energy efficient management system has become an essential concept to be considered while designing the system.

The future scope lies in the field of its application where these devices can play their role for more number of years and its sensing capability will be highly intelligent in architecture. In recent scenario, sensors are required that will be easily installed in human for health monitoring, military applications or environment activity monitoring of large geographical area. All these applications require low power consumption and advanced reliable technology. Also, different optimization techniques and their computational cost can be compared.

References

1. Hanes, D.: IoT fundamentals: networking technologies, protocols, and use cases for the internet of things. Cisco Press (2017)

2. Nolan, K.E., Guibene, W., Kelly, M.Y.: An evaluation of low power wide area network technologies for the Internet of Things. In: International Wireless Communications and Mobile Computing Conference (IWCMC), pp. 439–444. IEEE (2016)
3. Afzal, B., Umair, M.: Enabling IoT platforms for social IoT applications: vision, feature mapping, and challenges. Futur. Gener. Comput. Syst. **92**, 718–731 (2019)
4. Díaz, C.A.R.: IoT of: a long-reach fully passive low-rate upstream PHY for IoT over fiber. Electronics **8**(3), 359 (2019)
5. Al-Fuqaha, A., Guizani, M., Mohammadi, M., Aledhari, M., Ayyash, M.: Internet of things: a survey on enabling technologies, protocols, and applications. IEEE Commun. Surv. Tutor. **17**(4), 2347–2376 (2015)
6. Slabicki, M., Premsankar, G., Di Francesco, M.: Adaptive configuration of LoRa networks for dense IoT deployments. In: NOMS IEEE/IFIP Network Operations and Management Symposium. IEEE (2018)
7. Bor, M., Roedig, U.: LoRa transmission parameter selection. In: 13th International Conference on Distributed Computing in Sensor Systems (DCOSS). IEEE (2017)
8. Gautam, K., Puri, V., Tromp, J.G., Nguyen, N.G., Van Le, C.: Internet of Things (IoT) and deep neural network-based intelligent and conceptual model for smart city. In: Satapathy, S.C., Bhateja, V., Nguyen, B.L., Nguyen, N.G., Le, D.-N. (eds.) Frontiers in Intelligent Computing: Theory and Applications. AISC, vol. 1013, pp. 287–300. Springer, Singapore (2020). https://doi.org/10.1007/978-981-32-9186-7_30
9. Niknafs, M.: Neural network optimization. Technical report. https://courses.mai.liu.se/FU/MAI0083/Report_Mina_Nikanfs.pdf
10. Sandoval, R.M., Garcia-Sanchez, A.-J., Garcia-Haro, J.: Performance optimization of LoRa nodes for the future smart city/industry. EURASIP J. Wirel. Commun. Network. (2019)
11. Bouguera, T., Diouris, J.F., Chaillout, J.J., Jaouadi, R., Andrieux, G.: Energy consumption model for sensor nodes based on LoRa and LoRaWAN. Sensors **18**(7), 2104 (2018)
12. Sandoval, R.M., Garcia-Sanchez, A.-J., Garcia-Haro, J.: Optimizing and updating LoRa communication parameters: a machine learning approach. IEEE Trans. Netw. Serv. Manag. **16**(3), 884–895 (2019)
13. Kerkouche, R., Alami, R., Féraud, R., Varsier, N., Maillé, P.: Node-based optimization of LoRa transmissions with Multi-Armed Bandit algorithms. In: 25th International Conference on Telecommunications (ICT), pp. 521–526. IEEE (2018)
14. Hoeller, A., Souza, R.D., López, O.L.A., Alves, H., de Noronha Neto, M., Brante, G.: Analysis and performance optimization of LoRa networks with time and antenna diversity. IEEE Access **6**, 32820–32829 (2018)
15. Pasolini, G., et al.: Smart city pilot projects using LoRa and IEEE802.15.4 technologies. Sensors **18**(4), 1118 (2018)
16. Cuomo, F., Campo, M., Caponi, A., Bianchi, G., Rossini, G., Pisani, P.: EXPLoRa: extending the performance of LoRa by suitable spreading factor allocations. In: IEEE 13th International Conference on Wireless and Mobile Computing, Networking and Communications (WiMob), pp. 1–8. IEEE (2017)
17. Narieda, S., Fujii, T., Umebayashi, K.: Energy constrained optimization for spreading factor allocation in LoRaWAN. Sensors **20**(16), 4417 (2020)
18. Sanchez-Iborra, R., Sanchez-Gomez, J., Ballesta-Viñas, J., Cano, M.D., Skarmeta, A.F.: Performance evaluation of LoRa considering scenario conditions. Sensors **18**(3), 772 (2018)
19. Bor, M., Vidler, J.E., Roedig, U.: LoRa for the Internet of Things, pp. 361–366 (2016)
20. Verma, A., Prakash, S., Srivastava, V., Kumar, A., Mukhopadhyay, S.C.: Sensing, controlling, and IoT infrastructure in smart building: a review. IEEE Sens. J. **19**(20), 9036–9046 (2019)

21. Peura, U.-P.: LoRaWAN optimization for a battery powered sensor network. Theseus.fi (2018)
22. Liando, J.C., Gamage, A., Tengourtius, A.W., Li, M.: Known and unknown facts of LoRa: experiences from a large-scale measurement study. ACM Trans. Sens. Netw. (TOSN) **15**(2), 1–35 (2019)

Novel Hybrid GWO-WOA and BAT-PSO Algorithms for Solving Design Optimization Problems

Noel Jose Thengappurackal Laiju, Reza Sedaghat[✉],
and Prathap Siddavaatam

OPRA-Labs, Ryerson University, 350 Victoria Street,
Toronto, ON M5B 2K3, Canada
{nthengap,rsedagha,prathap.siddavaatam}@ee.ryerson.ca
https://www.ee.ryerson.ca/opr/

Abstract. This paper aims at the design and development of two hybrid nature inspired algorithms based on Grey Wolf Optimizer and Whale Optimization Algorithm (GWOWOA) and Binary Bat Optimization Algorithm and Particle Swarm Optimization Algorithm (BATPSO). Hybridization is a useful method to enhance the performance of these algorithms. The GWO algorithm is easy to fall into local optimum especially when it is used in the high-dimensional data. WOA algorithm experiences relatively low convergence precision and poor rate of convergence when it is applied in complex optimization problems. In this paper we embed bubble-net foraging method in WOA with that of prey encircling method in GWO. It has been observed experimentally that than just updating position vectors with respect to the three best fitness solutions, we were able to achieve faster convergence and better global optimum in most cases. In BATPSO, both algorithms are integrated and run in parallel and they perform a comparison between both minimum fitness function at each iteration. According to the observation, this greedy search algorithm in BAT optimization works best in higher values; however, finds it difficult in finding the global minimum as it reaches lower values; especially fractional fitness value. PSO is based on element-wise pos[i,j] search and updating the velocity to converge to a global minimum. The proposed hybrid algorithms are bench-marked using a set of 23 classical benchmark functions employed to test different characteristics of hybrid optimizers. The paper also performs solving two classical engineering design problems - Cantilever beam design and the multiple disc clutch brake problem. The results of the fitness functions prove that the proposed hybrid algorithms are able to produce better or very competitive output with respect to improved exploration, local optima avoidance, exploitation and convergence. All these hybrid algorithms find superior optimal designs for quintessential engineering problems engaged, showcasing that these algorithms are capable of solving constrained complex problems with diverse search spaces.

Keywords: Grey Wolf optimization · Whale Optimization Algorithm · Binary Bat Optimization Algorithm · Particle Swarm Optimization ·

© Springer-Verlag GmbH Germany, part of Springer Nature 2021
M. L. Gavrilova and C. J. K. Tan (Eds.): Trans. on Comput. Sci. XXXVIII, LNCS 12620, pp. 113–144, 2021.
https://doi.org/10.1007/978-3-662-63170-6_7

Design Space Exploration · Algorithm design and analysis ·
Classification algorithms · Mathematical model · Heuristic algorithms ·
Optimization · Feature extraction · Constrained optimization problems

1   Introduction

In the last two decades, evolutionary optimization algorithms have grabbed high
attention in terms of research and solving a wide range of real world optimization problems. Evolutionary algorithms are inspired by the physical phenomena,
animal behavior and other optimized methodologies and concepts already existing in nature which are developed and enhanced over the course of thousands
or even millions of years. An optimization process is initialized by creating a
set of random solutions. These initial solutions are then united, reallocated or
derived over a pre-determined number of steps termed as iterations. An algorithm becomes unique in terms of its characteristics in mixing, allocating or
evolving these initial solutions during the optimization process. Most of these
algorithms take advantages of stochastic operators which makes them unique
from deterministic approaches.

Meta-heuristic algorithms search for the global optimum in a search space
by creating one or more random solutions for a given problem Holland (1993).
Hence, these algorithms have following advantages: problem independence, evo-
lution independence, local minima evasion and natural optimization inspirations
makes these algorithms makes it simple and follow a general and common framework, which imparts us scope to improve these algorithms with hybridization.
Some of the most popular algorithms in this field used in this paper are: Grey
Wolf Optimizer (GWO) Mirjalili et al. (2014a), Whale Optimization Algorithm
(WOA) Mirjalili and Lewis (2016), Binary Bat Optimization Algorithm (BAT)
Mirjalili et al. (2014b) and Particle Swarm Optimization Algorithm (PSO) Sen-
gupta and Kachave (2017).

Optimization exists in almost every aspect right from engineering design,
business planning, computer networking and to even leisure travel. Our ultimate
goal is to optimize something such as quality, profit and time. In optimization of
a design, the design objective could be simply to minimize the cost of production
or to maximize the efficiency of production.

A hybrid algorithm is an algorithm that combines two or more other algorithms that solve the same problem, either choosing one (depending on the data),
or switching between them over the course of the algorithm. This is generally
done to combine desired features of each, so that the overall algorithm is better
than the individual components.

A "Hybrid algorithm" does not refer to simply combining multiple algorithms
to solve a different problem – many algorithms can be considered as combinations
of simpler pieces – but only to combining algorithms that solve the same problem,
but differ in other characteristics, notably performance.

Many researchers have attempted the use of hybridization in order to enhance
the performance of these algorithms. This paper proposes following two hybrid

algorithms: Hybrid Grey Wolf Optimizer and Whale Optimization Algorithm (GWOWOA) and Hybrid Binary Bat Optimization Algorithm and Particle Swarm Optimization Algorithm (BATPSO), Sect. 2 of this paper discusses literature review. Section 3 deals with the concept of hybridization and implementation of hybrid algorithms. Section 4 of this paper showcases bench-marking and results of 23 fitness functions used. Section 5 analyzes algorithm performance in two real world engineering design problem and Sect. 6 concludes the paper and discusses possible future scope and improvements.

1.1 Related Works

Meta-heuristic optimization algorithms have demonstrated their capability in finding acceptable solutions to many optimization problems through iterations and their stochastic behavior. Evolutionary optimization algorithms are constructed by mimicking naturally optimized behavior and traits in population initialization, candidate evaluation, termination condition, selecting method, recombination and mutation. Interestingly, most of these evolutionary algorithms share identical traits such as they usually spawn a random population and evaluate its fitness through an iterative approach. Most of these meta-heuristic algorithms divide the search space into two, namely, exploitation and exploration. However, due to the asymmetrical nature between exploration and exploitation combined with the stochastic attributes of these evolutionary algorithms, these algorithms often get entrapped in a local minima. We use hybridization in order to impart these algorithms a better ability in avoiding or escaping from local minima entrapment.

A hybrid algorithm is an algorithm that combines two or more other algorithms that solve the same problem, either choosing one (depending on the data), or switching between them over the course of the algorithm. This is generally done to combine desired features of each, so that the overall algorithm is better than the individual components. "Hybrid algorithm" does not refer to simply combining multiple algorithms to solve a different problem – many algorithms can be considered as combinations of simpler pieces – but only to combining algorithms that solve the same problem, but differ in other characteristics, notably performance.

Many researchers have attempted the use of hybridization in order to enhance the performance of these algorithms. In Xu et al. (2017) an improved GWO algorithm combined with Cuckoo Search (CS) is proposed. By introducing the global-search ability of CS into GWO to update its best three solutions that are alpha-wolf, beta-wolf and delta-wolf, the search ability of GWO is strengthened, and the local minima shortcoming of GWO is offset. The Jitkongchuen (2015) paper proposed hybrid differential evolution with grey wolf optimizer algorithm which introduces a new improved mutation schemes. In this algorithm, the control parameters are self-adapted by learning from previous evolutionary search. Beside, the grey wolf optimizer algorithm is used to enhance the crossover strategy. A proposed discrete GWO algorithm in which a random leader selection is performed, and the probability for the main leader to be selected increases at the detriment of the other leaders across iteration is discussed in Martin et al. (2018). In Arora et al. (2019), a hybrid GWO with CSA (Crow-Search Algorithm) is proposed which combines both the algorithms and are run in parallel to obtain improved results.

In Mafarja and Mirjalili (2017), two hybridization models are used to design different feature selection techniques based on Whale Optimization Algorithm (WOA). In the first model, Simulated Annealing (SA) algorithm is embedded in WOA algorithm, while it is used to improve the best solution found after each iteration of WOA algorithm in the second model. The goal of using SA here is to enhance the exploitation by searching the most promising regions located by WOA algorithm. In paper Kaveh and Rastegar Moghaddam (2017) which attempts to enhance the original formulation of the WOA by hybridizing it with some concepts of the colliding bodies optimization (CBO) in order to improve solution accuracy, reliability and convergence speed. In Arslan and Toz (2018), paper proposed a hybrid clustering algorithm that integrates Fuzzy C-Means (FCM) and Whale Optimization Algorithm (WOA) using the Chebshev distance function. According to their experimental results, hybridizing WOA, have improved FCM algorithms in clustering performance. A combination of PSO used for exploitation phase and WOA for exploration phase in uncertain environment is discussed in paper Trivedi et al. (2018). Analysis of competitive results obtained from PSO–WOA in the paper validates its effectiveness compared to standard PSO and WOA algorithm. In Korashy et al. (2019), GWO-WOA hybrid optimization algorithm is proposed where the proposed method enhances the exploitative phase of the WOA using a leadership hierarchy of the grey wolf optimizer (GWO) to find the best optimum solution. In Hudaib et al. (2018) GWO-WOA hybridization is realized by making combination between these two algorithms by employing the WOA algorithm to skip the limitation on the group team members and make it unlimited and take the top three highest solutions as GWO works. In Hachimi and Singh (2018) GWO-WOA hybridization is performed by using the spiral equation in Grey Wolf Optimizer Algorithm for balance between the exploitation and the exploration process in the new hybrid approach; and also applying this equation in the whole population in order to refrain from the premature convergence and trapping in local minima.

A hybrid Bat algorithm (BA) and Artificial Bee Colony (ABC) algorithm is discussed in Nguyen et al. (2014). The several worst individual of Bats in BA will be replaced with the better artificial agents in ABC algorithm after running every Ri iterations, and on the contrary, the poorer agents of ABC will be replacing with the better individual of BA. The proposed communication strategy provides the information flow for the bats to communicate in Bat algorithm with the agents in ABC algorithm. In Imane and Kamel (2016) a hybrid approach to embed bat algorithm with tabu search is implemented in order to improve the results obtained by applying a standard Bat algorithm. The algorithm transforms the phase of selection new solution in the standard bat algorithm by the procedure of tabu search. This paper Pravesjit (2016) proposes a hybrid bat algorithm with natural-inspired algorithms for continuous optimization problem. In its study, the proposed algorithm combines the reproduction step from weed algorithm and genetic algorithm. The reproduction step is applied to clone each bat population by fitness values and the genetic algorithm is applied in order to expand the population.

A hybrid Particle Swarm Optimization (PSO) and Ant Bee Colony (ABC) is explained in Gao (2018) aiming at the disadvantage of poor convergence performance of PSO. A variations of the particle swarm optimization - simulated annealing optimization technique (PSOSA) hybrid algorithm is depicted in Tharmalingam and Raahemifar (2012). The effects of initializing the particles strategically within the solution space along with the application of the SA algorithm to the hybrid algorithm at each iteration are explored in the paper. Genetic algorithm (GA) has been proved to be efficient for optimization problems. It contains four operators including coding, selection, crossover and mutation. Due to some drawbacks, it cannot be applied on all optimization problems. In Sharma and Singhal (2015) a hybrid form of GA is presented with particle swarm optimization algorithm to overcome these limitations. Aiming at the poor local search capability of Particle Swarm Optimization(PSO) algorithm, a hybrid particle swarm optimization algorithm is proposed in Changxing et al. (2017). Firstly, the population is initialized by tent chaotic map to improve the diversity of the initial population. In the evolution process, the tabu search strategy is adopted to improve algorithm convergence rate. Combining the chaos optimization strategy, this algorithm could jump out of local optimization and improve the local search ability. In Tawhid (2018) the decoupling of the velocity vectors of the bats and the particles are updated independently for both the particles according to the weighted combination of the personal and global best solutions and K-nearest neighbors is used as a classifier for feature selection.

2 Hybrid Grey Wolf Optimizer and Artificial Bee Colony Optimizer Algorithm (GWOWOA)

2.1 Background on Grey Wolf Optimizer

Grey Wolf Optimizer meta-heuristic algorithm is developed by Mirjalili et al. (2014a). Grey wolf (Canis lupus) belongs to Canidae family. Grey wolves are considered as apex predators, meaning that they are at the top of the food chain. Grey wolves mostly prefer to live in a pack. The group size is 5–12 on average. Of particular interest is that they have a very strict social dominant hierarchy.

The leaders are a male and a female, called alphas. The alpha is mostly responsible for making decisions about hunting, sleeping place, time to wake, and so on. The alpha's decisions are dictated to the pack. However, some kind of democratic behavior has also been observed, in which an alpha follows the other wolves in the pack. In gatherings, the entire pack acknowledges the alpha by holding their tails down. The alpha wolf is also called the dominant wolf since his/her orders should be followed by the pack Mech (1999). The alpha wolves are only allowed to mate in the pack. Interestingly, the alpha is not necessarily the strongest member of the pack but the best in terms of managing the pack. This shows that the organization and discipline of a pack is much more important than its strength.

The second level in the hierarchy of grey wolves is beta. The betas are subordinate wolves that help the alpha in decision-making or other pack activities. The beta wolf can be either male or female, and he/she is probably the best candidate to be the alpha in case one of the alpha wolves passes away or becomes very old. The beta wolf should respect the alpha, but commands the other lower-level wolves as well. It plays the role of an advisor to the alpha and discipliner for the pack. The beta reinforces the alpha's commands throughout the pack and gives feedback to the alpha.

The lowest ranking grey wolf is omega. The omega plays the role of scapegoat. Omega wolves always have to submit to all the other dominant wolves. They are the last wolves that are allowed to eat. It may seem the omega is not an important individual in the pack, but it has been observed that the whole pack face internal fighting and problems in case of losing the omega. This is due to the venting of violence and frustration of all wolves by the omega(s). This assists satisfying the entire pack and maintaining the dominance structure. In some cases the omega is also the babysitters in the pack.

If a wolf is not an alpha, beta, or omega, he/she is called subordinate (or delta in some references). Delta wolves have to submit to alphas and betas, but they dominate the omega. Scouts, sentinels, elders, hunters, and caretakers belong to this category. Scouts are responsible for watching the boundaries of the territory and warning the pack in case of any danger. Sentinels protect and guarantee the safety of the pack. Elders are the experienced wolves who used to be alpha or beta. Hunters help the alphas and betas when hunting prey and providing food for the pack. Finally, the caretakers are responsible for caring for the weak, ill, and wounded wolves in the pack Mirjalili et al. (2014a).

Grey Wolf Optimizer mimics the leadership hierarchy and hunting mechanism of gray wolves in nature. In designing GWO, we consider the fittest solution as the alpha (α), second and third best solutions are named beta (β) and delta (δ) respectively. The rest of the candidate solutions are considered to be omega (ω). The GWO optimization hunting mechanism is governed by α, β and δ. The ω wolves follow these three wolves. Faris et al. (2018).

The three main phases of grey wolves hunting mechanism is given as follows:

- Tracking, chasing and drawing near the prey.
- Pursuing, encircling and hassling the pray until it halts.
- Attacking towards the prey. Muro (2011)

This hunting technique and the social hierarchy of grey wolves are mathematically modeled in order to design GWO and perform optimization.

2.2 Background on Whale Optimization Algorithm Optimization

Whale Optimization Algorithm is developed by Mirjalili and Lewis (2016). Whales are fancy creatures. They are considered as the biggest mammals in the world. An adult whale can grow up to 30 m long and 180 t weight. There are 7 different main species of this giant mammal such killer, Minke, Sei, humpback, right, finback, and blue. Whales are mostly considered as predators. They never sleep because they have to breathe from the surface of oceans. In fact, half of the brain only sleeps. The interesting thing about the whales is that they are considered as highly intelligent animals with emotion.

According to Hof and Van Der Gucht (2006), whales have common cells in certain areas of their brains similar to those of human called spindle cells. These cells are responsible for judgment, emotions, and social behaviors in humans. In other words, the spindle cells make us distinct from other creatures. Whales have twice number of these cells than an adult human which is the main cause of their smartness. It has been proven that whale can think, learn, judge, communicate, and become even emotional as a human does, but obviously with a much lower level of smartness. It has been observed that whales (mostly killer whales) are able to develop their own dialect as well. Another interesting point is the social behavior of whales. They live alone or in groups. However, they are mostly observed in groups. Some of their species (killer whales for instance) can live in a family over their entire life period. One of the biggest baleen whales is humpback whales (Megaptera novaeangliae). An adult humpback whale is almost as size of a school bus. Their favorite prey are krill and small fish herds Watkins and Schevill (1979). The most interesting thing about the humpback whales is their special hunting method. This foraging behavior is called bubble-net feeding method Mirjalili and Lewis (2016). Humpback whales prefer to hunt school of krill or small fishes close to the surface. It has been observed that this foraging is done by creating distinctive bubbles along a circle or '9'-shaped path. Before 2011, this behavior was only investigated based on the observation from surface. However, Goldbogen et al. (2013) investigated this behavior utilizing tag sensors. They captured 300 tag-derived bubble-net feeding events of 9 individual humpback whales. They found two maneuvers associated with bubble and named them 'upward-spirals' and 'double- loops'.

In the former maneuver, humpback whales dive around 12 m down and then start to create bubble in a spiral shape around the prey and swim up toward the surface. The later maneuver includes three different stages: coral loop, lobtail, and capture loop. Detailed information about these behaviors can be found in Goldbogen et al. (2013). It is worth mentioning here that bubble-net feeding is a unique behavior that can only be observed in humpback whales. The spiral bubble-net feeding maneuver is mathematically modeled in order to perform WOA optimization.

The three main phases of Whale Optimization Algorithm are given below:

- Encircling prey.
- Bubble-net attacking method (exploitation phase).
- Search for prey (exploration phase)

2.3 Mathematical Modeling

Encircling prey method in WOA is almost as similar to tracking, chasing and drawing near the prey in GWO. Encircling prey in WOA is represented by following equations:

$$D = |C.\ \mathbf{X}*(t) - \mathbf{X}\ (t)| \tag{1}$$

$$\mathbf{X}(t + 1) = \mathbf{X}*(t) - \mathbf{A}\ .\ \mathbf{D} \tag{2}$$

where t denotes the current iteration, $\mathbf{A}$ and $\mathbf{C}$ are co-efficient vectors, $\mathbf{X}*$ is the position vector of best solution obtained so far and $\mathbf{X}$ indicates the position vector, $||$ is the absolute value and is an element -by- element multiplication.

Vectors $\mathbf{A}$ and $\mathbf{C}$ are calculated using below equations:

$$\mathbf{A} = 2\mathbf{a}.\mathbf{r}_1 - \mathbf{a} \tag{3}$$

$$\mathbf{C} = 2.\mathbf{r}_2 \tag{4}$$

We integrate the shrinking encircling mechanism of bubble-net attacking method in WOA to GWO prey chasing, tracking and encircling phase in this hybrid optimizer. The shrinking encircling mechanism of bubble-net attacking method is achieved by decreasing the value of $\mathbf{a}$ in Eq. 3. Note that the fluctuation range of $\mathbf{A}$ is also decreased by $\mathbf{a}$. Shrinking encircling mechanism is represented by the equation below:

$$\mathbf{X}(t + 1) = \mathbf{X}*(t) - \mathbf{A}\ .\ \mathbf{D} \tag{5}$$

We also adapt the spiral updating position of whale and prey to mimic the helix shaped movement of humpback whales as given below:

$$\mathbf{X}(t + 1) = \mathbf{D}'.\ e^{bl}.\cos{(2\pi l)} + \mathbf{X}*(t) \tag{6}$$

In search for prey (exploration phase); we update the position of a search agent in the exploration phase by selecting a search agent in a random fashion instead of choosing the search agent with the best solution obtained so far. This technique and $|A| > 1$ underline exploration and allow WOA to carry out a global search.

$$D = |C.\ \mathbf{X}_{\mathbf{rand}} - \mathbf{X}| \tag{7}$$

$$\mathbf{X}(t + 1) = \mathbf{X}_{\mathbf{rand}} - \mathbf{A}\ .\ \mathbf{D} \tag{8}$$

In GWO meta-heuristic algorithm, following equations are used:

$$D = |C.\ \mathbf{Xp}\ (t) - \mathbf{X}\ (t)| \tag{9}$$

$$\mathbf{X}\ (t + 1) = \mathbf{Xp}\ (t) - \mathbf{A}\ .\ \mathbf{D} \tag{10}$$

where t denotes the current iteration, $\mathbf{A}$ and $\mathbf{C}$ are co-efficient vectors, $\mathbf{Xp}\ (t)$ is the prey position vector and $\mathbf{X}$ indicates position of grey wolf. Vectors $\mathbf{A}$ and $\mathbf{C}$ are calculated using below equations:

$$\mathbf{A} = 2\mathbf{a}\ .\ \mathbf{r} - \mathbf{a} \tag{11}$$

$$\mathbf{C} = 2.\mathbf{r}_2 \tag{12}$$

Where **a** linearly decreases from to 2 to 0 and r_1 and r_2 are random vectors in [0,1]

We save the fittest solution as alpha. Solution beta is worse than alpha but higher than delta. After saving these three best solutions obtained so far and then we instruct the other search agents to update their positions according to the best search agents. For this, the following equations are used.

$$\mathbf{D}_\alpha = \mid \mathbf{C_1}\,\mathbf{X}_\alpha - \mathbf{X}\mid, \mathbf{D}_\beta = \mid \mathbf{C_2}\,\mathbf{X}_\beta - \mathbf{X}\mid, \mathbf{D}_\delta = \mid \mathbf{C_3}\,\mathbf{X}_\delta - \mathbf{X}\mid \tag{13}$$

$$\mathbf{X_1} = \mathbf{X}_\alpha - \mathbf{A_1}.(\mathbf{D}_\alpha), \mathbf{X_2} = \mathbf{X}_\beta - \mathbf{A_2}.(\mathbf{D}_\beta), \mathbf{X_3} = \mathbf{X}_\delta - \mathbf{A_3}.(\mathbf{D}_\delta) \tag{14}$$

$$\mathbf{X}_{(}\mathbf{t}+1) = \frac{\mathbf{X_1} + \mathbf{X_2} + \mathbf{X_3}}{3} \tag{15}$$

The whole idea of GWO-WOA hybridization is to impart both GWO and WOA characteristics in terms of both exploration and exploitation capability of the algorithm. In order to realize it, we amalgamate the following WOA Eqs. 1, 5, 6, 7 and 8 to modify and adapt to following GWO Eqs. 9, 10, 13 and 14. Since bubble-net foraging method is more sophisticated prey encircling method than just updating position vectors with respect to the three best fitness solutions, we were able to achieve faster convergence and better global optimum in most cases.

3   Hybrid Binary Bat Optimization Algorithm and Particle Swarm Optimization Algorithm (BATPSO)

3.1   Background on Binary Bat Optimization Algorithm

Binary Bat Algorithm (BAT) was proposed by Yang (2010) based on the echolocation behavior of bats. Bats are fascinating animals. They are the only mammals with wings and they also have advanced capability of echolocation. It is estimated that there are about 996 different species which account for up to 20% of all mammal species Toth and Parsons (2013). Their size ranges from the tiny bumblebee bat (of about 1.5 to 2 g) to the giant bats with wingspan of about 2 m and weight up to about 1 kg. Microbats typically have forearm length of about 2.2 to 11 cm. Most bats uses echolocation to a certain degree; among all the species, microbats are a famous example as microbats use echolocation extensively while mega-bats do not Gannon (2011). Most microbats are insectivores. Microbats use a type of sonar, called, echolocation, to detect prey, avoid obstacles, and locate their roosting crevices in the dark. These bats emit a very loud sound pulse and listen for the echo that bounces back from the surrounding objects. Their pulses vary in properties and can be correlated with their hunting strategies, depending on the species. Most bats use short, frequency-modulated signals to sweep through about an octave, while others more often use constant-frequency signals for echolocation. Their signal bandwidth varies depends on the species, and often increased by using more harmonics.

Algorithm 1: GWO-WOA Framework

Data: W_n–magnitude of the wolf pack(30), Iter$_{MAX}$ – total number of iterations(500) more than 0, UR – set of ordered m-tuples ($^j\mathfrak{a}_i$, $^j\mathfrak{a}_i$,..., $^m\mathfrak{a}_i$), $\mathfrak{C}_{\mathcal{R}_{aw}}$, $\mathfrak{C}_{\mathcal{R}_{ap}}$ and $\mathfrak{C}_{\mathcal{R}_{area}}$ – user defined constraints, p – probability WOA

Result: Optimal Grey wolf or Whale position within given Design Space.

1 **begin**

2 | Randomly initialize W_n number of candidate $\overrightarrow{X_i}$ Grey wolves

3 | Calculate the fitness of each search agent as X_α(*best search agent*), X_β (*second bestsearch agent*) and X_δ (*third best search agent*) wolves

4 | Set $\mathtt{t}:=0$

5 | **repeat**

6 | | **foreach** $\overrightarrow{X_i} \in$ *UR* **do**

7 | | | Update $\overrightarrow{X_i}$ using Eq. (14)

8 | | **end**

9 | | Compute Eqs. (11, 12) to update **A, C**

10 | | **forall the** *Wolves* $\overrightarrow{X_i}$ *of UR* **do**

11 | | | **if** $p < 0.5$ **then**

12 | | | | **if** $|A| < 1$ **then**

13 | | | | | newPositions $\leftarrow$ get positions using Eq. (1)

14 | | | | **else**

15 | | | | | Select random search agent X_{rand}

16 | | | | | Update the position of each search agent by Eq. 5

17 | | | | **end**

18 | | | **end**

19 | | **end**

20 | | **if** $p > 0.5$ **then**

21 | | | Select random search agent X_{rand}

22 | | | Update the position of each search agent by Eq. 5 and 6

23 | | | Re-initiate search iteration if no better solution found

24 | | **end**

25 | | Check if any search agent goes beyond the search space and amend it

26 | | Update α, A and C

27 | | Calculate the fitness of each search agents

28 | | Update α, β, and δ positions as premier 3 best solutions.

29 | | | $t \leftarrow t + 1$

30 | | **end**

31 | **until** $t \leq Iter_{MAX}$

32 | Choose `Optimal Grey Wolf-Whale` position

33 **end**

Though each pulse only lasts a few thousandths of a second (up to about 8 to 10 ms), however, it has a constant frequency which is usually in the region of 25 kHz to 150 kHz. The typical range of frequencies for most bat species are in the region between 25 kHz and 100 kHz, though some species can emit higher frequencies up to 150 kHz.

Each ultrasonic burst may last typically 5 to 20 ms, and microbats emit about 10 to 20 such sound bursts every second. When hunting for prey, the rate of pulse emission can be sped up to about 200 pulses per second when they fly near their prey. Such short sound bursts imply the fantastic ability of the signal processing power of bats. In fact, studies shows the integration time of the bat ear is typically about 300 to 400 μs.

Amazingly, the emitted pulse could be as loud as 110 dB, and, fortunately, they are in the ultrasonic region. The loudness also varies from the loudest when searching for prey and to a quieter base when homing towards the prey. The travelling range of such short pulses are typically a few metres, depending on the actual frequencies Toth and Parsons (2013).

Microbats can manage to avoid obstacles as small as thin human hairs. Studies show that microbats use the time delay from the emission and detection of the echo, the time difference between their two ears, and the loudness variations of the echoes to build up three dimensional scenario of the surrounding. They can detect the distance and orientation of the target, the type of prey, and even the moving speed of the prey such as small insects. Indeed, studies suggested that bats seem to be able to discriminate targets by the variations of the Doppler effect induced by the wing-flutter rates of the target insects Yang (2010). Obviously, some bats have good eyesight, and most bats also have very sensitive smell sense. In reality, they will use all the senses as a combination to maximize the efficient detection of prey and smooth navigation. However, here we are only interested in the echolocation and the associated behaviour. Such echolocation behaviour of microbats can be formulated in such a way that it can be associated with the objective function to be optimized, and this make it possible to formulate new optimization algorithms Yang (2010).

Bat Algorithm (BAT) was proposed by Xin-She Yang based on the echolocation behavior of bats. The capability of location of microbats is fascinating as these bats can find their prey and discriminate different types of insects. BAT algorithm is based on following assumptions.

- All bats use echolocation to sense distance, and they also 'know' the difference between food/prey and background barriers in some magical way.
- Bats fly randomly with velocity vi at position xi with a fixed frequency fmin, varying wavelength λ and loudness A0 to search for prey. They can automatically adjust the wavelength (or frequency) of their emitted pulses and adjust the rate of pulse emission $r \in [0, 1]$, depending on the proximity of their target.
- Although the loudness can vary in many ways, we assume that the loudness varies from a large (positive) A0 to a minimum constant value A_{min} [Yang(2010)]

Another obvious simplification is that no ray tracing is used in estimating the time delay and three dimensional topography. Though this might be a good feature for the application in computational geometry, however, we will not use this as it is more computationally extensive in multidimensional cases.

In addition to these simplified assumptions, we also use the following approximations, for simplicity. In general the frequency f in a range [fmin, fmax] corresponds to a range of wavelengths $[\gamma_{min}, \gamma_{max}]$. For example a frequency range of [20 kHz, 500 kHz] corresponds to a range of wavelengths from 0.7 mm to 17 mm.

For a given problem, we can also use any wavelength for the ease of implementation. In the actual implementation, we can adjust the range by adjusting the wavelengths (or frequencies), and the detectable range (or the largest wavelength) should be chosen.

3.2   Background on Particle Swarm Optimization

Particle swarm optimization (PSO) is a population based stochastic optimization technique developed by Dr. Eberhart and Dr. Kennedy in 1995, inspired by social behavior of bird flocking or fish schooling.

Imagine a flock of birds circling over an area where they can smell a hidden source of food. The one who is closest to the food chirps the loudest and the other birds swing around in his direction. If any of the other circling birds comes closer to the target than the first, it chirps louder and the others veer over toward him. This tightening pattern continues until one of the birds happens upon the food. It's an algorithm that's simple and easy to implement.

The algorithm keeps track of three global variables:

– Target value or condition.
– Global best (gBest) value indicating which particle's data is currently closest to the Target.
– Stopping value indicating when the algorithm should stop if the Target isn't found.[]

Each particle consists of:

– Data representing a possible solution.
– A Velocity value indicating how much the Data can be changed
– A personal best (pBest) value indicating the closest the particle's Data has ever come to the Target

The particles' data could be anything. In the flocking birds example above, the data would be the X, Y, Z coordinates of each bird. The individual coordinates of each bird would try to move closer to the coordinates of the bird which is closer to the food's coordinates (gBest). If the data is a pattern or sequence, then individual pieces of the data would be manipulated until the pattern matches the target pattern Eberhart and Kennedy (1995).

The velocity value is calculated according to how far an individual's data is from the target. The further it is, the larger the velocity value. In the birds example, the individuals furthest from the food would make an effort to keep up with the others by flying faster toward the gBest bird. If the data is a pattern or sequence, the velocity would describe how different the pattern is from the target, and thus, how much it needs to be changed to match the target.

Each particle's pBest value only indicates the closest the data has ever come to the target since the algorithm started. The gBest value only changes when any particle's pBest value comes closer to the target than gBest. Through each iteration of the algorithm, gBest gradually moves closer and closer to the target until one of the particles reaches the target Changxing et al. (2017).

3.3   Mathematical Modeling

In simulations, we use virtual bats naturally. We have to define the rules how their positions x_i and velocities v_i in a d-dimensional search space are updated. The new solutions x_i^t and velocities v_i^t at time step t are given by:

$$f_i = f_{min} + \beta(f_{max} - f_{min}) \tag{16}$$

$$v_i^t = v_i^{t-1} + (x_i^t - x*)f_i \tag{17}$$

$$x_i^t = x_i^{t-1} + v_i^t \tag{18}$$

where, $\beta \in [0,1]$, is a random vector drawn from a uniform distribution. Here x* is the current global best location (solution) which is located after comparing all the solutions amongst all the n bats.

As the product $\lambda_i f_i$ is the velocity increment, we can use either λ_i or f_i to adjust the velocity change while fixing the other factor λ_i or f_i, depending on the type of the problem of interest. In our implementation, we will use f_{min} = 0 and f_{max} = 100, depending the domain size of the problem of interest. Initially, each bat is randomly assigned a frequency which is drawn uniformly from $[f_{min}, f_{max}]$.

For the local search part, once a solution is selected among the current best solutions, a new solution for each bat is generated locally using random walk.

$$X_{new} = x_{old} + \in A^t \tag{19}$$

where $\mathcal{E} = [-1,1]$ is a random number, while $A_t = <A_{ti}>$ is the average loudness of all the bats at this time step.

The update of the velocities and positions of bats have some similarity to the procedure in the standard particle swarm optimization Kennedy (1995) as f_i essentially controls the pace and range of the movement of the swarming particles. To a degree, BAT can be considered as a balanced combination of the standard particle swarm optimization and the intensive local search controlled by the loudness and pulse rate. This is the core basis of our hybridization.

Furthermore, the loudness A_i and rate r_i of pulse emission is given by:

$$x_i^t = x_i^{t-1} + v_i^t \tag{20}$$

$$A_i^{t+1} = \alpha A_i^t \tag{21}$$

$$r_i^{t+1} = r_i^0[1 - e^{-\gamma t}] \tag{22}$$

where α and γ are constants. In fact, it is similar to the cooling factor of a cooling schedule in the simulated annealing.

PSO learned from the scenario and used it to solve the optimization problems. In PSO, each single solution is a "bird" in the search space. We call it "particle". All of particles have fitness values which are evaluated by the fitness function to be optimized, and have velocities which direct the flying of the particles. The particles fly through the problem space by following the current optimum particles.

PSO is initialized with a group of random particles (solutions) and then searches for optima by updating generations. In every iteration, each particle is updated by following two "best" values. The first one is the best solution (fitness) it has achieved so far. (The fitness value is also stored.) This value is called p_{best}. Another "best" value that is tracked by the particle swarm optimizer is the best value, obtained so far by any particle in the population. This best value is a global best and called g_{best}. When a particle takes part of the population as its topological neighbors, the best value is a local best and is called l_{best}.

In PSO, w parameter is given by below equation:

$$w = wMax - (current - iter) * ((wMax - wMin)/Max - iter) \tag{23}$$

where wMax is 0.9, wMin is 0.2. This w parameter is responsible for updating velocity in PSO given by:

$$\begin{aligned} v[] = w * v[] + c1 * rand() * (p_{best}[] - present[]) \\ + c2 * rand() * (g_{best}[] - present[]) \end{aligned} \tag{24}$$

$$present[] = present[] + v[] \tag{25}$$

$v[]$ is the particle velocity, present[] is the current particle (solution). $p_{best}[]$ and $g_{best}[]$ are defined as stated before. rand () is a random number between (0,1). c1, c2 are learning factors. Usually c1 = c2 = 2 Basturk (2006).

In BATPSO, In BAT minimum fitness function is located by an array-based greedy algorithm. All fitness function of array pos[i:] parameters are calculated and 'min' function is used to find the fitness minimum. According to the observation, this greedy algorithm works best in higher values; however, finds it difficult in finding the global minimum as it reaches lower values; especially fractional fitness value. PSO is based on element-wise pos[i,j] search and updating the velocity to converge to a global minimum. PSO works best at lower bound boundary fitness values. Therefore, in this hybrid, both PSO and BAT are run in parallel and they perform a comparison between both minimum fitness function at each iteration. The lowest value is taken and both PSO and BAT is updated with the lowest value and respective positions are updated simultaneously.

4   Experimental Study and Discussion

4.1   Fitness Functions and Parameter Settings

A fitness function is a particular type of objective function that is used to summarise, as a single figure of merit, how close a given design solution is to achieving the set aims. Fitness functions are used in genetic programming and genetic algorithms to guide simulations towards optimal design solutions. The test functions used in bench-marking are basically minimization functions and can be divided into four groups: unimodal, multimodal, fixed dimension multimodal and composite functions.

Unimodal functions given in Table 1 have only one global optimum. These functions allow to evaluate the exploitation capability of the investigated metaheuristic algorithms. In contrast to the unimodal functions, multimodal functions defined in Table 2 have many local optima with the number increasing exponentially with dimension. This makes them suitable for benchmarking the exploration ability of an algorithm. Unlike unimodal functions, fixed-dimension multimodal functions in Table 3 include many local optima whose number increases exponentially with the problem size (number of design variables). Therefore, this kind of test problems turns very useful if the purpose is to evaluate the exploration capability of an optimization algorithm. The last set of fitness functions in Table 4 is a combination of various constrained benchmark functions which are shifted, expanded, rotated and integrated in order to provide greater complexity in terms of testing the exploitation, exploration and effectiveness of optimization algorithms.

4.2   Graphical Analysis

A graphical analysis on hybrid algorithms with respect to their parent algorithms are examined in this section. The hybrid algorithm along with their parent algorithms are run simultaneously on 23 constrained bench-marking functions namely - unimodal, multimodal, fixed dimension multimodal and composite functions in order to graphically analyze the rate of convergence of specified algorithms and visually validate their performance with respect to their exploitation and exploration capability. All these algorithms are run 500 iterations and rate of convergence of all the algorithms are observed.

Algorithm 2: BAT-PSO Framework

Data: $x_i(i = 1, 2, \cdots, n)$ –number of bat population,
$v_i(i = 1, 2, \cdots, n)$ –velocity of bat population,
$d_i(i = 1, 2, \cdots, n)$ – number of PSO dimension
f_i – pulse frequency at x_i, r_i – pulse rate, A_i – loudness, total number of
iterations(500) more than 0
Result: Optimal BAT or PSO position within given Design Space

1 **begin**
2   initialize bat population $x_i(i = 1, 2, \cdots, n)$
3   and velocity $v_i(i = 1, 2, \cdots, n)$
4   Define pulse frequency f_i at x_i
5   Define pulse rate r_i and loudness A_i
6   Set **t**:=0
7   **repeat**
8   Generate new solutions by adjusting frequency,
9   and updating velocities and locations/solutions
10   equations 16 to 19.
11   **if** $rand > r_i$ **then**
12   Select a solution among the best solutions.
13   Generate a local solution around the selected best solution.
14   **end**
15   Generate a new solution by flying randomly ($Léwy flight$).
16   **if** $rand < A_i \& f(x_i) < f(x*)$ **then**
17   Accept the new solutions.
18   Increase r_i and reduce A_i
19   **end**
20   **foreach** *each dimension d indexed by i* **do**
21   Pick random numbers: $r_p, r_g \sim U(0, 1)$
22   Update the particle's velocity using Eq. 24
23   Update particle position using Eq. 25
24   **if** $f(x_i) < f(p_{best})$ **then**
25   Update $p_{best} \leftarrow x_i$
26   **end**
27   **if** $f(p_{best}) < f(g_{best})$ **then**
28   Update $g_{best} \leftarrow p_{best}$
29   **end**
30
31   **end**
32   Rank the bats and find the current best $x*$.
33   Compare PSO and BAT solution and update position.
34
35   **until** $t \leq Iter_{MAX}$
36   Choose **best** $PSO - -BAT$ **Postprocess results**
37 **end**

Table 1. Unimodal fitness functions

Function	Dim	Range	f_{min}
$f_1(x) = \sum_{i=1}^{n} x_i^2$	30	$[-100, 100]$	0
$f_2(x) = \sum_{i=1}^{n} |x_i| + \Pi_{i=1}^{n} |x_i|$	30	$[-10, 10]$	0
$f_3(x) = \sum_{i=1}^{n} (\sum_{j-1}^{i} x_j^2)$	30	$[-100, 100]$	0
$f_4(x) = max_i(|x_i|, 1 \le i \le n)$	30	$[-100, 100]$	0
$f_5(x) = \sum_{i=1}^{n-1} [100(x_{i+1} - x_i^2)^2 + (x_i - 1)^2]$	30	$[-30, 30]$	0
$f_6(x) = \sum_{i=1}^{n} [(x_i + 0.5)^2]$	30	$[-100, 100]$	0
$f_7(x) = \sum_{i=1}^{n} i x_i^4 + random[0, 1]]$	30	$[-1.28, 1.28]$	0

4.2.1 GWOWOA Graphical Analysis

Figure 1 shows the line graphs of convergence of GWOWOA hybrid algorithm
with respect to their parent GWO and WOA algorithms. From the graph, it
can be observed that GWOWOA has the best performance for this unimodal
benchmark function F_5. As unimodal functions allow to evaluate the exploitation
capability of the investigated meta-heuristic algorithms; GWOWOA algorithm
prove to have an improved optimum exploitation proficiency as it has very fast
convergence rate and constantly beats both GWO and WOA and achieve the best
final fitness value. GWO compete with GWOWOA until abouth 60th iteration
and then slows down and finally achieves an inferior output compared to both
WOA and GWOWOA. WOA got stuck in local minima until 200th iteration and
then it converges steeply beating GWO and ended up accomplishing a slightly
better result than GWO.

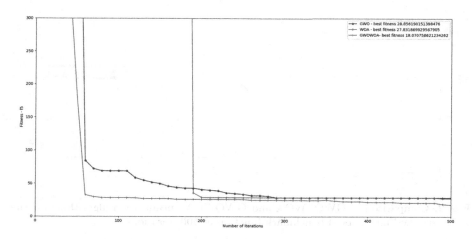

Fig. 1. Comparison of GWO, WOA and GWOWOA optimization algorithms for the
f5 constrained unimodal benchmark function in 500 iterations.

Table 2. Multimodal fitness functions

Function	Dim	Range	f_{min}
$f_8(x) = \sum_{i=1}^{n} -x_i \sin(\sqrt{|x_i|})$	30	$[-500, 500]$	-418.9289 x5
$f_9(x) = \sum_{i=1}^{n} [x_i^2 - 10\cos(2\pi x_i) + 10$	30	$[-5.12, 5.12]$	0
$f_{10}(x) = -20\exp -0.2\sqrt{\frac{1}{n}\sum_{i=1}^{n} x_i^2} -$ $\exp(1n\sum_{i=1}^{n}\cos(2\pi x_i)) + 20 + e$	30	$[-32, 32]$	0
$f_{11}(x) = \frac{1}{4000}\sum_{i=1}^{n} x_i^2 - \Pi_{i=1}^{n} cos(\frac{x_i}{\sqrt{i}}) + 1)$	30	$[-600, 600]$	0
$f_{12}(x) = 0.1[\sin^2(3\pi x_i) + \sum_{i=1}^{n}(x_i - 1)^2$ $+[1 + \sin(3\pi x_i + 1] + (x_n - 1)^2 + [1 + \sin^2(2\pi x_n)]]$ $+\sum_{i=1}^{n} u(x_i, 5, 100, 4)$	30	$[-50, 50]$	0
$f_{13}(x) = \sum_{i=1}^{n} sin(x_i).(\sin(\frac{i.x_i^2}{\pi}))^{2m}, m = 10$	30	$[0, \Pi]$	-4.687

Table 3. Fixed-dimension multimodal fitness functions

Function	Dim	Range	f_{min}
$f_{14}(x) = (\frac{1}{500}\sum_{j=1}^{25} \frac{1}{j+\sum_{i=1}^{2}(x_i - a_{ij})^6})^{-1}$	2	$[-65, 65]$	1
$f_{15}(x) = \sum_{i=1}^{11}(a_i - \frac{x_1(b_i^2 + b_i x_2)}{b_i^2 + b_i x_3 + x_4})^2$	4	$[-5, 5]$	0.00030
$f_{16}(x) = 4x_1^2 + 2.1x_1^4 + \frac{1}{3}x_1^6 - 4x_2^2 + x_1 x_2 + 4x_2^4)$	2	$[-5, 5]$	-1.0316
$f_{17}(x) = \sum_{i=1}^{4} c_i \exp(-\sum_{j=1}^{3} a_{ij}(x_j - p_{ij})^2)$	2	$[1, 3]$	-3.86
$f_{18}(x) = \sum_{i=1}^{5}[t(X - a_i)(X - a_i t)^T + c_i]^{-1}$	4	$[0, 10]$	-10.1532
$f_{19}(x) = \sum_{i=1}^{10}[t(X - a_i)(X - a_i t)^T + c_i]^{-1}$	4	$[0, 10]$	-10.5363

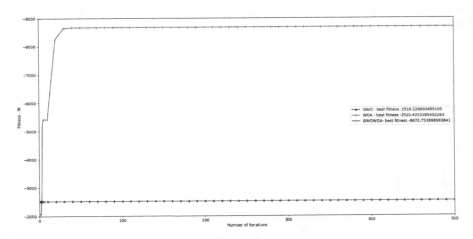

Fig. 2. Comparison of GWO, WOA and GWOWOA optimization algorithms for the f_8 constrained multimodal benchmark function in 500 iterations.

Figure 2 shows the line graphs of convergence of GWOWOA hybrid algorithm with respect to their parent GWO and WOA algorithms. From the graph, it can be observed that GWOWOA has the best performance for this multimodal

Table 4. Composite fitness functions

Function	Dim	Range	f_{min}
f_{20}(CF1) : $f_1, f_2, f_3, \ f_4, \ \dots, f_{10} = Sphere \ function$ $\beta_1, \beta_2, \beta_3, \dots, \beta_{10}] = [1, 1, 1, \dots, 1]$ $[\lambda_1, \lambda_2, \lambda_3, \dots, \lambda_{10} = [5/100, 5/100, 5/100, \dots, 5/100]$	10	$[-5, 5]$	0
f_{21}(CF2): $f_1, f_2, f_3, \ f_4, \ \dots, f_{10} = Griewank's \ function$ $\beta_1, \beta_2, \beta_3, \dots, \beta_{10}] = [1, 1, 1, \dots, 1]$ $[\lambda_1, \lambda_2, \lambda_3, \dots, \lambda_{10} = [5/100, 5/100, 5/100, \dots, 5/100]$	10	$[-5, 5]$	0
f_{22}(CF3): $f_1, f_2 = Ackley's function$ $f_3, \ f_4, \ = Rastrigin's \ function$ $f_5, \ f_6, \ = Weierstra's \ function$ $f_7, \ f_8, \ = Griewank's \ function$ $f_9, \ f_{10}, \ = Sphere \ function$ $\beta_1, \beta_2, \beta_3, \dots, \beta_{10}] = [1, 1, 1, \dots, 1]$ $[\lambda_1, \lambda_2, \lambda_3, \dots, \lambda_{10} = [[5/32, 5/32, 1, 1, 5/0.5, 5/0.5, 5/100, 5/100, 5/100, 5/100]$	10	$[-5, 5]$	0
f_{23}(CF4) : $f_1, f_2 = Ackley's function$ $f_3, f_4 = Rastrigin's \ function$ $f_5, f_6 = Weierstra's \ function$ $f_7, f_8 = Griewank's \ function$ $f_9, f_{10} = Sphere \ function$ $\beta_1, \beta_2, \beta_3, \dots, \beta_{10}] = [1, 1, 1, \dots, 1]$ $[\lambda_1, \lambda_2, \lambda_3, \dots, \lambda_{10} = [0.1 * 1/5, 0.2 * 1/5, 0.3 * 5/0.5, 0.4 * 5/0.5, 0.5 * 5/100,$ $0.6 * 5/100, 0.7 * 5/32, 0.8 * 5/32, 0.9 * 5/100, 1 * 5/100]$	10	$[-5, 5]$	0

benchmark function F_8. Multiimodal functions allow to evaluate the exploration capability of the investigated meta-heuristic algorithms. Exploration consists of probing a much larger portion of the search space with the hope of finding other promising solutions that are yet to be refined. This operation amounts then to diversifying the search in order to avoid getting trapped in a local optimum. The graphical analysis clearly shows that GWOWOA showcases immaculate exploration capability in order to jump outside from a local optimum. From the graph, we can infer that both GWO and WOA are entrapped in local optimum and improve negligibly with subsequent iterations. Whereas, GWOWOA displays excellent potential in jumping out from local minima at multiple occasions and finally acquire a significantly better global optimum.

Figure 3 demonstrates rate of convergence on GWOWOA, GWO and WOA algorithm in fixed dimension multimodal fitness function f_{15}. Fixed dimension multimodal systems have complex search space due the presence of multiple local minima. Therefore these constrained benchmark functions test exploration ability of an optimization algorithm in a more complex search space. From the graph it is evident that GWO tends to fall in local minima at lower values eventhough it had the fastest convergence at early stage. WOA stagnates at local minima until around 160th iteration and reaches a better global optimum than GWO but at a slower convergence rate. GWOWOA converges very slowly

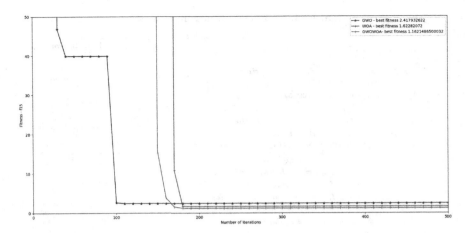

Fig. 3. Comparison of GWO, WOA and GWOWOA optimization algorithms for the f_{15} constrained fixed dimensional multimodal benchmark function in 500 iterations.

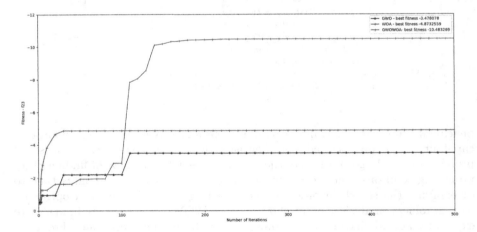

Fig. 4. Comparison of GWO, WOA and GWOWOA optimization algorithms for the f_{23} constrained composite benchmark function in 500 iterations.

initially and surpass GWO and maintains better global optimum than both GWO and WOA in subsequent iterations.

Figure 4 illustrates the convergence rate in composite constrained benchmark function F_{23}. Composite benchmark functions investigate both exploitation and exploration ability of an optimizer. The convergence line graph shows that GWO-WOA have very acute convergence rate towards lower fitness values and demonstrates better local minima avoidance at lower fitness values. It starts very slowly and outpace GWO at 80th iteration and WOA around 100th iteration and achieves a superior global optimum value compared to both GWO and WOA.

4.2.2   BATPSO Graphical Analysis

Figure 5 shows the line graphs of convergence of BATPSO hybrid algorithm with respect to their parent BAT and PSO algorithms. From the graph, it can be observed that initially BATPSO have the worst initial fitness function value compared to BAT and PSO. However, BATPSO exhibits rapid convergence speed and surpasses PSO and BAT before 50th iteration and arrive at better global optimum than both BAT and PSO in this unimodal benchmark function F_4. Unimodal fitness functions allow to evaluate the exploitation capability of the investigated meta-heuristic algorithms. PSO achieves a better fitness value at the end compared to that of BAT algorithm.

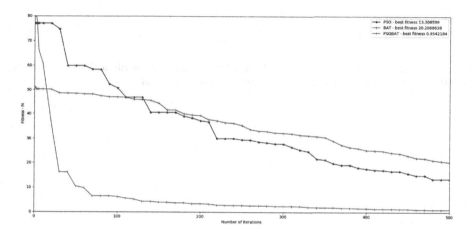

Fig. 5. Comparison of BATPSO, BAT and PSO optimization algorithms for the f_4 constrained unimodal benchmark function in 500 iterations.

Figure 6 shows the line graphs of convergence of BATPSO hybrid algorithm with respect to their parent BAT and PSO algorithms. From the graph, it can be observed that BATPSO has the fastest convergence rate compared to PSO and BAT for this multimodal benchmark function F_9. Multimodal functions allow to evaluate the exploration capability of the investigated meta-heuristic algorithms. BATPSO beats PSO algorithm around 20th iteration and surpass BAT at 50th iteration. BATPSO arrives at a superior global optimum at later iterations and displays better exploration characteristics than both BAT and PSO. Figure 7 demonstrates rate of convergence on BATPSO, MVO and PSO algorithms in fixed dimension multimodal fitness function f_{17}. Fixed dimension multimodal systems have complex search space due the presence of multiple local minima. Therefore these constrained benchmark functions test exploration ability of an optimization algorithm in a more complex search space. From the graph it is observed that PSO is the fastest and has sharp convergence rate in the beginning. BAT has the slowest convergence graph and very bottom out result compared to PSO and BATPSO. BATPSO arrives at improved global

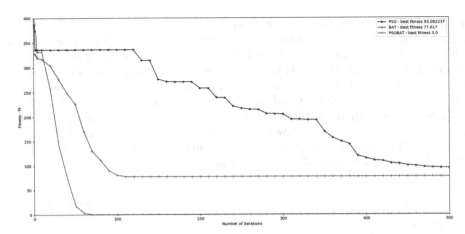

Fig. 6. Comparison of BATPSO, BAT and PSO optimization algorithms for the f_9 constrained multimodal benchmark function in 500 iterations.

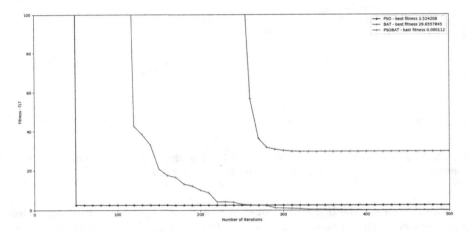

Fig. 7. Comparison of BATPSO, BAT and PSO optimization algorithms for the f_{17} constrained fixed dimension multimodal benchmark function in 500 iterations.

minimum than the rest at 300th iteration and have a better optimum in subsequent iterations. Figure 8 illustrates the convergence rate in composite constrained benchmark function F_{23}. Composite benchmark functions investigate both exploitation and exploration ability of an optimizer. The convergence line graph shows that BATPSO have faster convergence rate than PSO and BAT. BAT stagnates at local minimum and then jump out of it and merges with PSO at 280th iteration. BATPSO maintain steady and superior rate of convergence and demonstrates superior global optimum throughout 500 iterations.

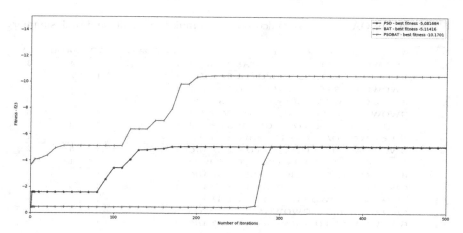

Fig. 8. Comparison of BATPSO, BAT and PSO optimization algorithms for the f_{23} constrained composite benchmark function in 500 iterations.

4.3   Statistical Analysis

Statistical testing is a process of making quantitative decisions about a problem in which statistical data set is evaluated and taken which is then compared hypothetically Wilcoxon et al. (1963). In this analysis we consider that there are instances in which is very essential to carry out analysis along the search.

This way, optimization algorithms can be evaluated depending on its convergence performance, which would help when deciding which algorithms perform better among a set of methods that are assumed as equal when only the results at the end of the search are considered. In addition to hybrid algorithms and their parent algorithms, two hybrid evolutionary algorithms such as Hybrid Bat algorithm and Artificial Bee Colony Algorithm (BATABC) Nguyen et al. (2014) and Memetic Algorithm (MA) Moscato (1999) is also considered in order to validate the performance of hybrid algorithm realized.

4.3.1   Wilcoxon Signed Rank Test

The Wilcoxon signed rank test is a non parametric test which means the population data is non-normally distributed. It is solely based on the order of the sample's observations Wilcoxon et al. (1963). The one with lowest rank will be considered as the best amongst all and vice-versa.

The results of Wilcoxon signed rank test for GWOWOA is given in Table 5. The results shows that GWOWOA have the lowest rank among all five algorithms subjected to the statistical testing. Their parent algorithms GWO and WOA competed with GWOWOA and was ranked second and third respectively. BATABC and MA algorithms shared the fourth place.

The results of Wilcoxon signed rank test for BATPSO is given in Table 6. The results shows that BATPSO have the lowest rank among all five algorithms

Table 5. GWOWOA - Pair-wise Wilcoxon signed rank test results and rank summary

Function	Wilcoxon signed rank test order	GWOWOA	GWO	WOA	BATABC	MA
f_1	$\mathbf{GWOWOA} < GWO < MA < BATABC < WOA$	1	2	5	4	3
f_2	$\mathbf{GWOWOA} < BATABC < GWO < WOA < MA$	1	3	4	2	5
f_3	$\mathbf{GWOWOA} < BATABC < MA < WOA < GWO$	1	5	4	2	3
f_4	$WOA < \mathbf{GWOWOA} < MA < GWO < BATABC$	2	4	1	5	3
f_5	$\mathbf{GWOWOA} < WOA < GWO < BATABC < MA$	1	3	2	4	5
f_6	$GWO < BATABC < WOA < \mathbf{GWOWOA} < MA$	4	1	3	2	5
f_7	$MA < \mathbf{GWOWOA} < BATABC < WOA < GWO$	2	5	4	3	1
f_8	$\mathbf{GWOWOA} < MA < WOA < GWO < BATABC$	1	4	3	5	2
f_9	$BATABC < MA < GWO < \mathbf{GWOWOA} < WOA$	4	3	5	1	2
f_{10}	$WOA < \mathbf{GWOWOA} < GWO < MA < BATABC$	2	3	1	5	4
f_{11}	$\mathbf{GWOWOA} < WOA < GWO < MA < BATABC$	1	3	2	5	4
f_{12}	$GWO < MA < \mathbf{GWOWOA} < WOA < BATABC$	3	1	4	5	2
f_{13}	$GWO < BATABC < \mathbf{GWOWOA} < MA < WOA$	3	1	5	2	4
f_{14}	$\mathbf{GWOWOA} < GWO < MA < WOA < BATABC$	1	2	4	5	3
f_{15}	$GWO < \mathbf{GWOWOA} < WOA < BATABC < MA$	2	1	3	4	5
f_{16}	$BATABC < WOA < GWO < \mathbf{GWOWOA} < MA$	4	3	2	1	5
f_{17}	$BATABC < WOA < MA < GWO < \mathbf{GWOWOA}$	5	4	2	1	3
f_{18}	$WOA < \mathbf{GWOWOA} < MA < BATABC < GWO$	2	5	1	4	3
f_{19}	$\mathbf{GWOWOA} < GWO < WOA < MA < BATABC$	1	2	3	5	4
f_{20}	$MA < WOA < GWO < BATABC < \mathbf{GWOWOA}$	5	3	2	4	1
f_{21}	$\mathbf{GWOWOA} < MA < GWO < BATABC < WOA$	1	3	5	4	2
f_{22}	$\mathbf{GWOWOA} < GWO < BATABC < MA < WOA$	1	2	5	3	4
f_{23}	$BATABC < \mathbf{GWOWOA} < WOA < GWO < MA$	2	4	3	1	5
Total		50	71	73	75	75

Table 6. BATPSO - Pair-wise Wilcoxon signed rank test results and rank summary

Function	Wilcoxon signed rank test order	BATPSO	PSO	BAT	BATABC	MA
f_1	$\mathbf{BATPSO} < PSO < BATABC < BAT < MA$	1	2	4	3	5
f_2	$BATABC < \mathbf{BATPSO} < PSO < BAT < MA$	2	3	4	1	5
f_3	$\mathbf{BATPSO} < MA < PSO < BAT < BATABC$	1	3	4	5	2
f_4	$MA < PSO < \mathbf{BATPSO} < BATABC < BAT$	3	2	5	4	1
f_5	$MA < BATABC < BAT < PSO < \mathbf{BATPSO}$	5	4	3	2	1
f_6	$\mathbf{BATPSO} < PSO < MA < BATABC < BAT$	1	2	5	4	3
f_7	$BATABC < PSO < MA < \mathbf{BATPSO} < BAT$	4	2	5	1	3
f_8	$PSO < MA < BATABC < BAT < \mathbf{BATPSO}$	5	1	4	3	2
f_9	$MA < \mathbf{BATPSO} < BAT < BATABC < PSO$	2	5	3	4	1
f_{10}	$\mathbf{BATPSO} < PSO < MA < BATABC < BAT$	1	2	5	4	3
f_{11}	$BAT < \mathbf{BATPSO} < PSO < MA < BATABC$	2	3	1	5	4
f_{12}	$MA < BATABC < \mathbf{BATPSO} < BAT < PSO$	3	5	4	2	1
f_{13}	$\mathbf{BATPSO} < PSO < BAT < MA < BATABC$	1	2	3	5	4
f_{14}	$MA < \mathbf{BATPSO} < BAT < BATABC < PSO$	2	5	3	4	1
f_{15}	$\mathbf{BATPSO} < BAT < PSO < MA < BATABC$	1	3	2	5	4
f_{16}	$\mathbf{BATPSO} < MA < BAT < BATABC < PSO$	1	5	3	4	2
f_{17}	$BATABC < \mathbf{BATPSO} < PSO < BAT < MA$	2	3	4	1	5
f_{18}	$BAT < MA < PSO < BATABC < \mathbf{BATPSO}$	5	3	1	4	2
f_{19}	$PSO < \mathbf{BATPSO} < BATABC < BAT < MA$	2	1	4	3	5
f_{20}	$MA < BAT < PSO < BATABC < \mathbf{BATPSO}$	5	3	2	4	1
f_{21}	$BATABC < \mathbf{BATPSO} < BAT < PSO < MA$	2	4	3	1	5
f_{22}	$\mathbf{BATPSO} < PSO < BATABC < BAT < MA$	1	2	4	3	5
f_{23}	$PSO < \mathbf{BATPSO} < BATABC < MA < BAT$	2	1	5	3	4
Total		54	66	81	75	69

Table 7. GWOWOA - Comparison results of all constrained benchmark functions

Function	GWOWOA	GWO	WOA	BATABC	MA
f_1	0.7718	**0.4336**	2.7718	2.0119	2.9911
f_2	**0.3118**	0.8465	0.7611	1.3312	**0.3311**
f_3	**34.3399**	40.6651	**34.5511**	88.2237	55.7718
f_4	**3.6691**	6.745	4.1182	9.5518	11.0063
f_5	53.221	**48.5418**	66.7182	**48.9918**	201.987
f_6	**0.4001**	1.182	**0.3888**	5.331	1.9012
f_7	**0.4512**	**0.04531**	0.8129	0.9318	1.3128
f_8	−8813.26	**−9321.85**	**−9331.11**	−4418.89	−6318.22
f_9	**48.599**	51.612	67.829	58.631	**50.129**
f_{10}	**1.4918**	**1.4254**	3.118	**1.4671**	7.691
f_{11}	0.5161	**0.3622**	0.8819	**0.3812**	0.9991
f_{12}	**0.4412**	0.9812	0.9812	**0.4618**	1.6311
f_{13}	**0.0541**	0.0416	1.778	1.116	**0.05885**
f_{14}	1.1121	**0.9980**	**0.9980**	**0.9980**	1.1221
f_{15}	**0.1629**	1.2318	**0.1911**	0.6719	0.8916
f_{16}	**−1.0413**	**−1.0337**	**−1.0408**	−1.2981	−6.5211
f_{17}	0.7621	1.562	**0.4421**	**0.4611**	1.1222
f_{18}	**2.819**	**2.998**	**2.979**	3.122	**2.999**
f_{19}	−2.8125	**−3.8621**	−1.8712	**−3.8821**	**−3.8015**
f_{20}	**−3.3618**	**−3.3525**	−3.1128	−2.9912	−1.3697
f_{21}	**−10.078**	−7.8912	−3.8187	**−10.1019**	−3.6981
f_{22}	**−10.1381**	**−10.1244**	−7.8121	−8.1281	−2.8911
f_{23}	**−10.8218**	−10.6365	**−10.8715**	−9.5211	−8.1269

subjected to the statistical testing. PSO competed closely with BATPSO and was ranked second. MA finished at third position at a considerable recession point. BAT algorithm was the worst performer and its hybrid version BATABC was ranked fourth. The superior performance of GWOWOA and BATPSO does not imply it is better than the other algorithms selected for the case study which will lead to the violation of 'free lunch theorem' Ho and Pepyne (2002). Its performance outlines its better performance in selected benchmark functions in this literature.

4.3.2   Mean-Average Method

This is the simplest and most popular method used for statistical testing of optimization algorithm. In order to have an accurate evaluation of hybrid algorithm, all algorithms selected for statistical testing is run 30 times. Then we take the mean average of each optimization algorithm in order to validate the results.

Table 8. BATPSO - Comparison results of all constrained benchmark functions

Function	BATPSO	BAT	PSO	BATABC	MA
f_1	**0.6152**	2.1123	**0.6652**	2.0119	2.9911
f_2	**0.3815**	0.4565	0.4186	1.3312	**0.3311**
f_3	39.5561	40.6651	**34.0012**	88.2237	55.7718
f_4	**3.7196**	6.745	**3.6936**	9.5518	11.0063
f_5	53.221	56.1125	66.7182	**48.9918**	201.987
f_6	**0.3699**	**0.3816**	**0.3771**	5.331	1.9012
f_7	**0.4401**	0.9128	0.8129	0.9318	1.3128
f_8	**−9332.18**	−5019.12	**−9338.67**	−4418.89	−6318.22
f_9	57.192	51.612	67.829	58.631	**50.129**
f_{10}	**1.4396**	1.6178	3.118	**1.4671**	7.691
f_{11}	0.5161	**0.3881**	1.1108	**0.3812**	0.9991
f_{12}	**0.4561**	0.9812	**0.4506**	**0.4618**	1.6311
f_{13}	**0.0561**	0.0889	0.5819	1.116	**0.0588**
f_{14}	**0.9980**	**0.9980**	**0.9980**	**0.9980**	1.1221
f_{15}	1.6012	**0.1887**	**0.1805**	0.6719	0.8916
f_{16}	**−1.0511**	**−1.0537**	−1.3316	−1.2981	−6.5211
f_{17}	0.4351	2.0169	**0.4881**	**0.4611**	1.1222
f_{18}	3.0167	**2.998**	**2.998**	3.122	**2.999**
f_{19}	**−3.8799**	**−3.8881**	−1.8712	**−3.8821**	**−3.8015**
f_{20}	**−3.3771**	−2.0018	−3.1128	−2.9912	−1.3697
f_{21}	**−10.078**	−7.8912	−3.8187	**−10.1019**	−3.6981
f_{22}	−9.6511	**−10.1301**	−7.8121	−8.1281	−2.8911
f_{23}	**−10.7018**	−6.8915	**−10.7715**	−9.5211	−8.1269

Comparison results of GWOWOA and BATPSO are given in Table 7 and Table 8. According to Table 7, GWOWOA were able to find better solutions in 16 out of 23 benchmark functions than the other selected optimization algorithms. GWO provided competitive results in almost every benchmark function, however, it edge past GWOWOA results in only seven benchmark functions and tied in 5 benchmark functions. WOA also performed well and output the best results in 9 benchmark functions. FA and DA algorithms gave fair results but they were significantly low results compared to that of GWOWOA and GWO. As per Table 8, BATPSO generated best global optimum in 16 out of 23 benchmark functions. PSO gave very competitive results in almost all benchmark functions ad surpassed all other algorithms in 8 benchmark functions and shared optimum values in 4 benchmark functions. BAT algorithm provided very good results and gave best results in 8 fitness functions. BATABC algorithm performed well giving best global optimum in 8 fitness functions. MA algorithm performed lower than the other selected algorithms producing best results in 5 fitness functions only.

5   Constrained Optimization and Classical Engineering problems

Engineering design is the method that engineers use to identify and solve problems. In constrained engineering design process, engineers must identify solutions that include the most desired features and fewest negative characteristics. They should also specify the cost functions and their limitations of the given scenario, which could include time, cost, and the physical limits of tools and materials. Constrained engineering design optimization problems are usually computationally expensive due to non-linearity and non convexity of these constraint functions. Evolutionary population based algorithms are widely used to solve constrained optimization problems. Many researchers have implemented many heuristic and meta-heuristic optimization algorithms to solve constrained optimization problems in engineering design.

These meta-heuristic optimization algorithms are of great research interest in recent times due to their ability in finding optimal solutions within short time especially when these real world engineering design problems consists of large number of design variables and multiple constraints which makes the solution search-space larger, complicated and non-linear. Penalty function methods are found to be quite popular due to their simplicity and ease of implementation. In this method, search agents are assigned big objective function values if they violate any of the specified constraints. In this section, we try to solve a real world engineering design problem using hybrid algorithm in order to observe the performance and benchmark the performance when compared to its parent and other popular evolutionary algorithms.

5.1   Cantilever Beam Design Using GWOWOA

Table 9. Multiple disc clutch brake problem results using BATPSO

Parameters	BATPSO	PSO	BAT	BATABC	MA
z_1	70	70	70	70	70
z_2	90	90	90	90	90
z_3	1	1	1	1	1.5
z_4	860	910	910	910	980
z_5	3	3	3	3	3
$f(\mathbf{z})$(Optimum Weight)	0.31367	0.31367	0.31367	0.31367	0.42651

This is a structural optimization problem Seyedali Mirjalili (2015). The objective is to design a minimum-mass cantilever beam. A cantilever beam includes five

Table 10. Cantilever Beam Design results using GWOWOA

Parameters	GWOWOA	GWO	WOA	BATABC	MA
z_1	6.0089	6.0100	6.0100	6.0089	6.0152
z_2	5.3081	5.3000	5.3044	5.3051	5.3000
z_3	4.4900	4.4900	4.5011	4.4970	4.4900
z_4	3.4950	3.4980	3.4900	3.4968	3.5033
z_3	2.1512	2.1500	2.1503	2.1504	2.1556
$f(\mathbf{z})$(Optimum Weight)	1.3399	1.3400	1.34006	1.34009	1.34207

hollow elements with square-shaped cross-section. Since the mass is proportional to the cross-sectional area of the beam, the objective function for the problem is taken as the cross-sectional area. Assuming thickness is constant, there are a total of 5 structural parameters. The mathematical formulation of this problem can be described as follows:

$$Consider,$$
$$\mathbf{z} = [z_1 z_2 z_3 z_4 z_5],$$
$$Minimize\ the\ function,$$
$$f(\mathbf{z}) = 0.6224(z_1 + z_2 + z_3 + z_4 + z_5),$$
$$Subject\ to, \tag{26}$$
$$h_1(\mathbf{z}) = \frac{61}{z_1^3} + \frac{37}{z_2^3} + \frac{19}{z_3^3} + \frac{4}{z_4^3} + \frac{1}{z_5^3} \leq 1$$
$$where.$$
$$0.01 \leq z_1, z_2, z_3, z_4, z_5 \leq 100,$$

Table 10 compares the best solutions for Cantilever Beam Design problem obtained by the proposed GWOWOA, its parent optimizers GWO and WOA as well as BATABC and MA optimization algorithm. As per the result GWOWOA was able to find the optimal solution $f(\mathbf{z})$ of 1.3399 which is very competitive to its parent algorithms. This evidences that the proposed algorithm is able to effectively optimize challenging constrained problems as well. All algorithms had obeyed all constraints. BATABC and MA algorithm also provided very emulous results.

5.2   Multiple Disc Clutch Brake Problem Using BATPSO

The objective of this optimization problem is to minimize the mass of multiple disc clutch brake. The design problem is constituted of five variables namely inner radius, outer radius and thickness of the disc, actuating force and the number of friction surfaces Osyczka and Krenich (2004). They are denoted as $z_1, z_2, ..., z_5$. The problem is subjected to a total of 8 constraints. Mathematical formulation of this problem is given below:

$$Minimize\ the\ function,$$
$$f(\mathbf{z}) = \pi(r_o^2 - r_i^2)t(Z+1)p,$$

$$Subject\ to,$$
$$h_1(\mathbf{z}) = r_o - r_i - \Delta r \geq 0,$$
$$h_2(\mathbf{z}) = l_{max} - (Z+1)(t+\delta) \geq 0,$$
$$h_3(\mathbf{z}) = p_{max} - p_{rz} \geq 0,$$
$$h_4(\mathbf{z}) = p_{max}v_{st\ max} - p_{rz}v_{st} \geq 0,$$
$$h_5(\mathbf{z}) = v_{st\ max} - v_{st} \geq 0,$$
$$h_6(\mathbf{z}) = T_{max} - T \geq 0,$$
$$h_7(\mathbf{z}) = M_h - sM_s \geq 0,$$
$$h_8(\mathbf{z}) = T \geq 0,$$

$$(27)$$

$$where.$$

$$\mu = 0.5, I_z = 55\,\mathrm{kgmm}^2, n = 250\,\mathrm{rpm}, \Delta r = 20\,\mathrm{mm}, M_s = 40\,\mathrm{Nm},$$
$$p_{max} = 1MP_a, T_{max} = 15\,\mathrm{s}, F_{max} = 1000\,\mathrm{N}, s = 1.5, M_f = 3\,\mathrm{Nm},$$
$$v_{st\ max} = 10, I_{max} = 30\,\mathrm{mm}, r_{i\ min} = 60, r_{o\ min} = 90, r_{o\ max} = 110,$$
$$r_{i\ max} = 80, t_{max} = 3, F_{min} = 600, Z_{min} = 2, Z_{max} = 9, t_{min} = 1.5$$

$$All\ variables\ are\ discrete\ and\ have\ following\ values,$$
$$z_1 = 60, 61, \cdots, 80;$$
$$z_2 = 90, 91, \cdots, 110;$$
$$z_3 = 1, 1.5, \cdots, 3;$$
$$z_4 = 600, 610, \cdots, 1000;$$
$$z_5 = 2, 3, \cdots, 9;$$

Table 9 give the best solutions attained by the selected optimization algorithms along with BATPSO. BATPSO was able to attain the best solution; very slightly improved solution compared to that of MVO and PSO. FA and DE were outperformed by BATPSO, MVO and PSO algorithms.

6   Conclusion

This paper proposed two new hybrid algorithms, GWOWOA and BATPSO. GWOWOA was inspired from Grey Wolf Optimization (GWO) and Whale Optimization Algorithm (WOA). BATPSO was based on Binary Bat Optimization Algorithm (BAT) and Particle Swarm Optimization (PSO) Algorithm. Hybridization improved the exploration and exploitation ability of optimization algorithms. The proposed algorithms were successfully graphically analyzed for performance, rate of convergence and local minima avoidance on 23 different classical constrained benchmark functions. These 23 benchmark fitness functions composed of unimodal, multimodal, fixed-dimension multimodal and composite fitness functions in order to evaluate exploration and exploitation ability of optimization algorithms effectively.

Three different statistical testing were carried out to validate the results. The results outlined the superiority of the proposed GWOWOA and BATPSO algorithms in solving numerous constrained and unconstrained problems. In addition, the attained results show that proposed GWOWOA and BATPSO acquire significantly improved solutions compared with their parent and other popular optimization algorithms. Moreover, we used these proposed hybrid optimization algorithms to solve two real world engineering design problems with large number of variables and constraints. The results show the capability of GWOWOA and BATPSO algorithms in handling various real world conjunctional optimization problems under lower computational efforts. Both GWOWOA and BATPSO were able to attain the optimal or near optimal solutions better than to most of the existing optimization algorithms subjected to the study.

7   Future Research

For future work, large scale optimization problems can be studied and realized using these algorithms. A self-adaptive method of choosing parameters can be developed in order to further improve the efficiency of these optimization algorithms. A multi-objective version of these algorithms can also be developed which find itself immense scope in diverse real world optimization applications.

References

Watkins, W.A., Schevill, W.E.: Aerial observation of feeding behavior in four baleen whales: Eubalaena glacialis, balaenoptera borealis, megaptera novaeangliae, and balaenoptera physalus. J. Mammalogy **60**, 155–159 (1979)

Arora, S., Singh, H., Sharma, M., Sharma, S., Anand, P.: A new hybrid algorithm based on grey wolf optimization and crow search algorithm for unconstrained function optimization and feature selection. IEEE Access 1–6 (2019)

Arslan, H., Toz, M.: Hybrid FCM-WOA data clustering algorithm. In: 2018 26th Signal Processing and Communications Applications Conference (SIU), pp. 1–4, May 2018

Basturk, B., Karaboga, D.: An artificial bee colony (ABC) algorithm for numeric function optimization. In: IEEE Swarm Intelligence Symposium, pp. 4–12 (2006)

Changxing, Q., Yiming, B., Huihua, H., Yong, L.: A hybrid particle swarm optimization algorithm. In: 2017 3rd IEEE International Conference on Computer and Communications (ICCC), pp. 2187–2190, December 2017

Eberhart, R.C., Kennedy, J.: A new optimizer using particle swarm theory. In: Proceedings of the Sixth International Symposium on Micro Machine and Human Science, pp. 39–43 (1995)

Faris, H., Aljarah, I., Al-Betar, M.A., Mirjalili, S.: Grey wolf optimizer: a review of recent variants and applications. Neural Comput. Appl. **30**(2), 413–435 (2018)

Gannon, W.L., Kunz, T.H., Parsons, S. (eds.): Ecological and Behavioral Methods for the Study of Bats, 2nd edn., p. 901. Johns Hopkins University Press, Baltimore (2009). ISBN 978-0-8018-9147-2, price (hardbound). J. Mammal. **92**(2), 475–478 (2011)

Gao, Y.: An improved hybrid group intelligent algorithm based on artificial bee colony and particle swarm optimization. In: 2018 International Conference on Virtual Reality and Intelligent Systems (ICVRIS), pp. 160–163, August 2018

Goldbogen, J.A., Friedlaender, A.S., Calambokidis, J., McKenna, M.F., Simon, M., Nowacek, D.P.: Integrative approaches to the study of baleen whale diving behavior, feeding performance, and foraging ecology. BioScience **63**, 90–100 (2013)

Hachimi, H., Singh, D.N.: A new hybrid whale optimizer algorithm with mean strategy of grey wolf optimizer for global optimization. Math. Comput. Appl. **23** (2018)

Ho, Y.C., Pepyne, D.L.: Simple explanation of the no free lunch theorem of optimization. Cybern. Syst. Anal. **38**(2), 292–298 (2002)

Hof, P.R., der Gucht, E.V.: Structure of the cerebral cortex of the humpback whale, megaptera novaeangliae (cetacea, mysticeti, balaenopteridae). Anat. Rec. **290**(1), 1–31 (2006)

Holland, J.H.: Genetic Algorithms, vol. 4. Sci Am, Chichester (1993)

Hudaib, A., Masadeh, R., Alzaqebah, A.: WGW: a hybrid approach based on whale and grey wolf optimization algorithms for requirements prioritization. Adv. Syst. Sci. Appl. **02**, 63–83 (2018)

Imane, M., Kamel, N.: Hybrid bat algorithm for overlapping community detection. IFAC-PapersOnLine **49**, 1454–1459 (2016)

Jitkongchuen, D.: A hybrid differential evolution with grey wolf optimizer for continuous global optimization. In: 2015 7th International Conference on Information Technology and Electrical Engineering (ICITEE), pp. 51–54, October 2015

Kaveh, A., Rastegar Moghaddam, M.: A hybrid WOA-CBO algorithm for construction site layout planning problem. Scientia Iranica **25** (2017)

Kennedy, J., Eberhart, R.: Particle swarm optimization. In: Proceedings of IEEE International Conference on Neural Networks, pp. 1942–1948, September 1995

Korashy, A., Kamel, S., Jurado, F., Youssef, A.R.: Hybrid whale optimization algorithm and grey wolf optimizer algorithm for optimal coordination of direction overcurrent relays. Electric Power Components Syst. **47**, 644–658 (2019)

Mafarja, M., Mirjalili, S.: Hybrid whale optimization algorithm with simulated annealing for feature selection. Neurocomputing (2017)

Martin, B., Marot, J., Bourennane, S.: Improved discrete grey wolf optimizer. In: 2018 26th European Signal Processing Conference (EUSIPCO), pp. 494–498, September 2018

Mech, L.D.: Alpha status, dominance, and division of labor in wolf packs. Can. J. Zool. **77**(8), 1196–1203 (1999)

Mirjalili, S., Lewis, A.: The whale optimization algorithm. Adv. Eng. Softw. **95**, 51–67 (2016)

Mirjalili, S., Mirjalili, S.M., Lewis, A.: Grey wolf optimizer. Adv. Eng. Softw. **69**, 46–61 (2014a)

Mirjalili, S., Mirjalili, S.M., Yang, X.S.: Binary bat algorithm. Neural Comput. Appl. **25**(3–4), 663–681 (2014). https://doi.org/10.1007/s00521-013-1525-5

Tawhid, M.A., Dsouza, K.B.D.: Hybrid binary dragonfly enhanced particle swarm optimization algorithm for solving feature selection problems. Math. Found. Comput. **01**, 181 (2018)

Moscato, P.: Memetic algorithms: a short introduction. In: Corne, D., et al. (eds.) New Ideas in Optimization, pp. 219–234. McGraw-Hill Ltd., Maidenhead (1999)

Muro, C., Escobedo, R., Spector, L., Coppinger, R.P.: Wolf-pack (canis lupus) hunting, strategies emerge from simple rules in computational simulations. Behav. Process **88**, 92–99 (2011)

Nguyen, T.T., Pan, J.S., Dao, T.K., Kuo, M.Y., Horng, M.F.: Hybrid bat algorithm with artificial bee colony. In: Pan, J.S., Snasel, V., Corchado, E.S., Abraham, A., Wang, S.L. (eds.) Intelligent Data analysis and its Applications, vol. II, pp. 45–55. Springer, Cham (2014). https://doi.org/10.1007/978-3-319-07773-4_5

Osyczka, A., Krenich, S.: Some methods for multicriteria design optimization using evolutionary algorithms. J. Theor. Appl. Mech. **42** (2004)

Pravesjit, S.: A hybrid bat algorithm with natural-inspired algorithms for continuous optimization problem. Artif. Life Robot. **21**(1), 112–119 (2016)

Sengupta, A., Kachave, D.: Particle swarm optimization driven low cost single event transient fault secured design during architectural synthesis (invited paper). J. Eng. **1**, 5 (2017)

Mirjalili, S., Mirjalili, S.M., Hatamlou, A.: Multi-verse optimizer: a nature-inspired algorithm for global optimization. Nat. Comput. Appl. Forum **10**, 15–21 (2015)

Sharma, J., Singhal, R.S.: Comparative research on genetic algorithm, particle swarm optimization and hybrid GA-PSO. In: 2015 2nd International Conference on Computing for Sustainable Global Development (INDIACom), pp. 110–114 (2015)

Tharmalingam, M., Raahemifar, K.: Strategic iniitialization of a hybrid particle swarm optimization-simullated annealing algorithm (HPSOSA) for PID controller design for a nonlinear system. In: 2012 25th IEEE Canadian Conference on Electrical and Computer Engineering (CCECE), pp. 1–4 (2012)

Toth, C., Parsons, S.: Is lek breeding rare in bats? J. Zool. **291**, 23–27 (2013)

Trivedi, I.N., Jangir, P., Kumar, A., Jangir, N., Totlani, R.: A novel hybrid PSO-WOA algorithm for global numerical functions optimization. In: Bhatia, S.K., Mishra, K.K., Tiwari, S., Singh, V.K. (eds.) Advances in Computer and Computational Sciences, pp. 53–60. Springer, Singapore (2018). https://doi.org/10.1007/978-981-10-3773-3_6

Wilcoxon, F., Katti, S., Wilcox, R.A.: Critical Values and Probability Levels for the Wilcoxon Rank Sum Test and the Wilcoxon Signed Rank Test. American Cyanamid, Pearl River (1963)

Xu, H., Liu, X., Su, J.: An improved grey wolf optimizer algorithm integrated with cuckoo search. In: 2017 9th IEEE International Conference on Intelligent Data Acquisition and Advanced Computing Systems: Technology and Applications (IDAACS), vol. 1, pp. 490–493, September 2017

Yang, X.S.: A New Metaheuristic Bat-Inspired Algorithm, pp. 65–74. Springer, Heidelberg (2010). https://doi.org/10.1007/978-3-642-12538-6_6

Author Index

Printed in the United States
by Baker & Taylor Publisher Services